'A remarkable book that provides a unique perspective on modern software development. A distinctive and unusual feature is the way modern software development principles are explained holistically in terms of all project activities. The focus on key employment skills and knowledge also makes it a must read for aspiring developers.'

Chris Beaumont PhD FBCS FHEA, *Chair of Examiners, NCC Education*

'I wish this book had been around when I was starting out 30 years ago. It's a manual for all aspects of software development and the scope of the role in business, rather than focusing on being a "coder". I particularly like the fact it includes client aspects, which are usually forgotten!'

Andy Doyle, *Director, Nice Group (SW) Ltd*

'As a leader of many software development teams, this book will be indispensable to modern developers and managers alike. It will not teach you how to write .net, but will help when someone who does tries to bamboozle you with jargon. It is brilliantly written and easy to digest.'

Paul Leonard, *Group Technology & Infrastructure Manager, DCC plc*

'A comprehensive, practical overview of what awaits you in the real world of professional software development.'

Karl Beecher, *Author of* Computational Thinking

'*Software Development in Practice* takes the guesswork out of your journey into tech. From term definitions to Agile practices and clean code tips, this book is my go-to resource for anyone breaking into the tech industry. I especially appreciate the emphasis on communication, collaboration and user experience.'

Sjoukje Ijlstra, *Software Engineer, JP Morgan*

'There are many books which describe various technical and theoretical aspects of software development. However, few describe what's actually involved in day-to-day software development. This book is one of those few and should be of real interest to prospective and early-career software developers.'

Dr Patrick Hill, *R&D Director, QPC Ltd*

'As a security researcher and advocate for embedding security in the software development process, it is enlightening to see this book dedicate some detailed coverage to consider use of defensive coding techniques, GDPR from a developer's point of view, and specific vulnerabilities and associated mitigations taken direct from the OWASP Top10.'

Adrian Winckles, *Director of Cyber and Networking, Anglia Ruskin University*

'A great book for those thinking of working or progressing in the commercial software development industry. The book gives insight into working practices, identifying positives and negatives to each of them. Deliberately avoiding specific programming languages (other than to explain some points), the book will be a perfect addition for any dev team in any development environment.'

Martin Thorne, *Technical Director, Montpellier Integrated*

'This book provides the framework to apply knowledge of how to code into the real world of being a software developer. It is the theory and thought processes that you can't learn without doing the job first – until now! If you're considering a career path in software development this book should be the first port of call on your journey.'

Kieran Purdie, *Pro AV Channel Manager & Business Development / Technical Manager, NETGEAR Business, UK & Ireland*

'If you want a guide on what you need to do to become a fantastic software developer, then this book is for you. The book's in-depth topic coverage will provide you with all the tools and information you will need to succeed in the software development Industry.'

Anthony Davis, *Senior Manager Platform Engineering, Sixt*

'IT now permeates almost every area of business. In an environment where the pace is ever increasing, it is essential for those aspiring to work as a software developer to gain knowledge, skills and experience in many areas. *Software Development in Practice* covers the areas to master to become a productive member of a software development team.'

Chris Galley FBCS CITP

SOFTWARE DEVELOPMENT IN PRACTICE

BCS, THE CHARTERED INSTITUTE FOR IT

BCS, The Chartered Institute for IT, is committed to making IT good for society. We use the power of our network to bring about positive, tangible change. We champion the global IT profession and the interests of individuals, engaged in that profession, for the benefit of all.

Exchanging IT expertise and knowledge
The Institute fosters links between experts from industry, academia and business to promote new thinking, education and knowledge sharing.

Supporting practitioners
Through continuing professional development and a series of respected IT qualifications, the Institute seeks to promote professional practice tuned to the demands of business. It provides practical support and information services to its members and volunteer communities around the world.

Setting standards and frameworks
The Institute collaborates with government, industry and relevant bodies to establish good working practices, codes of conduct, skills frameworks and common standards. It also offers a range of consultancy services to employers to help them adopt best practice.

Become a member
Over 70,000 people including students, teachers, professionals and practitioners enjoy the benefits of BCS membership. These include access to an international community, invitations to a roster of local and national events, career development tools and a quarterly thought-leadership magazine. Visit www.bcs.org/membership to find out more.

Further information
BCS, The Chartered Institute for IT,
3 Newbridge Square,
Swindon, SN1 1BY, United Kingdom.
T +44 (0) 1793 417 417
(Monday to Friday, 09:00 to 17:00 UK time)
www.bcs.org/contact
http://shop.bcs.org/

SOFTWARE DEVELOPMENT IN PRACTICE

Bernie Fishpool and Mark Fishpool

Published by BCS Learning & Development Ltd, a wholly owned subsidiary of BCS, The Chartered Institute for IT, 3 Newbridge Square, Swindon, SN1 1BY, UK.
www.bcs.org

Paperback ISBN: 978-1-78017-497-6
PDF ISBN: 978-1-78017-498-3
ePUB ISBN: 978-1-78017-499-0
Kindle ISBN: 978-1-78017-500-3

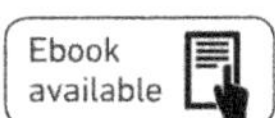

British Cataloguing in Publication Data.
A CIP catalogue record for this book is available at the British Library.

Publisher's acknowledgements
Reviewers: Patrick Hill, Karl Beecher
Publisher: Ian Borthwick
Commissioning editor: Rebecca Youé
Production manager: Florence Leroy
Project manager: Hazel Bird
Copy-editor: Hazel Bird
Proofreader: Barbara Eastman
Indexer: Sally Roots
Cover design: Alex Wright
Cover image: Shutterstock/nofilm2011
Typeset by Lapiz Digital Services, Chennai, India.

CONTENTS

LIST OF FIGURES AND TABLES

FIGURES

TABLES

AUTHORS

Bernie Fishpool is a consultant subject matter expert and author of vocational computing and IT texts and learning resources, who taught programming, systems analysis, project management and personal skills development in post-16 education for more than 20 years. On leaving teaching in 2007, she worked for Edexcel as a curriculum development manager. Bernie led the development of the AQA Tech-levels in IT between 2014 and 2016 as a sector strategist for IT and computing, and has written many Level 2 and Level 3 BTEC IT units for Pearson Edexcel over the years. More recently Bernie was a member of the team that developed the BCS Trailblazer Apprenticeships, and she has recently worked on the BCS Essential Digital Skills Qualification (EDSQ). She continues to develop teaching and learning resources.

Mark Fishpool has taught various computing subjects from Year 7 to degree level in schools and colleges since the age of 19. He has lead-written nationally recognised BTEC and AQA computing qualifications and was head of school for computing and Centre of Vocational Excellence manager at Gloucestershire College for over seven successful years before returning to the commercial sector, specialising as a senior full-stack developer. He has recently come back to training and education as senior technical learning consultant (cyber) at QA Ltd, specialising in Python, C and Linux. Somehow in the middle of this 30-year career he's also written 11 books, mostly with his wife, Bernie.

ABBREVIATIONS

ACID	atomicity, consistency, isolation and durability
API	application programming interface
AppSec	application security
AWS	Amazon Web Services
BIDMAS	brackets, indices, division, multiplication, addition, subtraction
BODMAS	brackets, orders, division, multiplication, addition, subtraction
CIW	Certified Internet Web Professional
CLI	command line interface
CMM	Capability Maturity Model
CMS	content management system
CPD	continuing professional development
CPU	central processing unit
CRUD	create, read, update, delete
CSS	Cascading Style Sheets
CSV	Comma-Separated Values
DCL	Data Control Language
DDaT	Digital, Data and Technology Professions
DDL	Data Definition Language
DevOps	developer and operations
DevSecOps	developer, security and operations
DFD	data flow diagram
DML	Data Manipulation Language
DQL	Data Query Language
DRY	don't repeat yourself
EAFP	easier to ask for forgiveness than it is to get permission
EC2	Elastic Compute Cloud (Amazon)
EE	Enterprise Edition (Java)
ERD	entity relationship diagram
EU	European Union
FAQs	frequently asked questions

FE	further education
FIFO	first in, first out
FTP	File Transfer Protocol
GDPR	General Data Protection Regulation
GUI	graphical user interface
HNC	Higher National Certificate
HND	Higher National Diploma
HTML	Hypertext Mark-up Language
HTTP	Hypertext Transfer Protocol
I/O	input and output
IaaS	infrastructure as a service
IAM	Identity and Access Management (Amazon)
IDE	integrated development environment
IoT	Internet of Things
JSON	JavaScript Object Notation
LAMP	Linux, Apache, MySQL and PHP
LBYL	look before you leap
LIFO	last in, first out
LMGTFY!	let me Google that for you!
MAMP	Mac, Apache, MySQL and PHP
MCA	Microsoft Certified Associate
MCSA	Microsoft Certified System Administrator
MCSD	Microsoft Certified Solutions Developer
MCSE	Microsoft Certified Solutions Expert
MVC	model-view-controller
MVP	minimum viable product
NoSQL	Not only SQL
OCA	Oracle Certified Associate
OCM	Oracle Certified Master
OCP	Oracle Certified Professional
OOP	object-oriented programming
OS	operating system
OTS	off-the-shelf
OWASP	Open Web Application Security Project
PaaS	platform as a service
PCAP	Certified Associate in Python Programming
PCEP	Certified Entry-Level Python Programmer
PCPP	Certified Professional in Python Programming
PEP8	Python Enhancement Proposal 8

PYPL	Popularity of Programming Language
QAT	quality assurance testing
RAD	rapid application development
RAM	random access memory
RDBMS	relational database management system
RESTful	representational state transfer
ROI	return on investment
SaaS	software as a service
SDLC	software development lifecycle
SE	Standard Edition (Java)
SFIA	Skills Framework for the Information Age
SHA1	Secure Hash Algorithm 1
SMS	short message service
SQL	Structured Query Language
SRP	single responsibility principle
TDD	test-driven development
UI	user interface
UML	Unified Modelling Language
UX	user experience
VCS	version control system
VM	virtual machine
WAMP	Windows, Apache, MySQL and PHP
WET	write everything twice (or we enjoy typing)
XML	Extensible Mark-up Language
XP	extreme programming
XXE	XML external entity
YAML	YAML Ain't Mark-up Language

PREFACE

> 'A software engineer, a developer, and coder walk into a bar.
> "Here come the programmers!", says the bartender.'
>
> – Kim (2018)

This book covers good practice and gives you some things to consider no matter where you are in the software development lifecycle.

From a historical perspective, programmers traditionally wrote code – that was their job. They didn't always get involved in analysis or testing, and they certainly wouldn't have been expected to have in-depth knowledge of hardware, networking or databases. That's perhaps a gross simplification, but the kernel of truth remains: software development could be a very isolated – and insular – activity.

The reality today is that for most software developers their job is **not just about coding**. They are expected to have a much wider field of expertise and almost certainly have an extensive list of essential and desirable IT skills, whether these relate to industry-trending code frameworks, modern methodologies, or experience with particular operating systems and popular IDE (integrated development environment). Having a working knowledge of a single programming language just doesn't cut the mustard any more.

For example, a commercial online organisation might require you to demonstrate 'full-stack' web development skills. This would mean knowing the principles of data design, database development, and server- and client-side interaction (e.g. JavaScript); HTML5 (Hypertext Mark-up Language version 5) and CSS3 (Cascading Style Sheets version 3); email and FTP (File Transfer Protocol) server integration; and much more. On the other hand, if you were developing for custom hardware, you might need to understand different signal types in Linux and couple this with advanced low-level C skills. Although both examples would require the skills of a software developer, the roles themselves, the underpinning knowledge and the daily challenges would be quite different.

What's clear is that the role of a software developer is a moving feast, and individuals need to continually improve and adapt their skills over time. This is especially the case for contract programmers as they move between commercially diverse projects.

In this role you'll find that learning never stops. To fill gaps in your own skill set, you may need to take additional courses or gain new professional qualifications to help you to prepare for this type of role. Above all, in these information-sensitive times, you should care about data protection principles and the role of cybersecurity and defensive coding.

A push in the late 1980s led the industry to embrace the notion of software engineering principles rather than simply using the term 'programmer' to define development, and this has now become the norm. The profession has become somewhat more formalised, more professional and far less 'wild west' – although, in some development environments, that pioneering spirit still plays a crucial part in thinking outside the proverbial box and creating amazing innovations.

Despite the passing of time, key concepts remain important: generality, consistency, incremental development, anticipation of change, abstraction, modularity, and the concepts 'separation of concerns', 'the single responsibility principle' and 'don't repeat yourself'. While rapidly evolving technology produces improved frameworks and encourages better workflows that enable teams to seamlessly share development assets, the underlying challenge of producing robust, reliable and – increasingly important – secure solutions within (typically) tight timescales steadfastly remains.

Many graduate programmers leave education with a sound theoretical understanding, having been taught using a series of tried-and-tested problems that can be resolved using well-chosen tools and techniques in a prescribed way to produce sensible (and somewhat predictable) solutions. Unfortunately, the real world is much messier than this, and few problems encountered 'in the wild' will decompose quite so neatly. This doesn't mean the textbooks should be thrown out or that the educational journey was not worthwhile (far from it), but there should be an acceptance that there is always going to be a great deal still to learn. And, of course, this is no bad thing.

For many new developers, consideration of the well-established software development lifecycle (see Chapter 4) can be a good place to start because it instils structure and process. However, as you will see, there are many different approaches and development models in vogue at any particular time, as they go in and out of fashion or are adapted and remixed for the next generation.

This book will help you to explore the workings of software development and will furnish you with the tools and understanding to make informed decisions as a software developer.

Ultimately, software development should be based on a series of fundamental principles that help developers to successfully navigate software projects. Which ones you choose to follow are up to you – software development is a broad church, and **everyone** is welcome.

WHO THIS BOOK IS FOR

This book is intended for those who aspire to work in a software development role within the IT sector.

It will be of interest to you if you are thinking of starting a software development career or have completed initial education and training programmes, such as A-levels, a Pearson BTEC, T-Levels or a Level 3 apprenticeship and are contemplating your next step. Of course, it will also be relevant to those who have worked for a while in a related or unrelated sector and are simply contemplating a career change.

Whatever your reason for considering a career in software development, you will need to understand the context of the profession, explore the end-to-end process involved, and get an idea of the good and bad practices, the tools and techniques, and the myriad knotty problems and rock-hard challenges that lie ahead.

Although written from an insider's perspective, this book has (hopefully) not lost the outsider's excitement of looking through a window into a world that appears to be tremendous fun, nose pressed against the glass, wondering how best to get involved. So, whether you consider yourself a budding software engineer, a developer or just a coder – welcome, programmers and aspiring programmers all!

1 GETTING STARTED IN SOFTWARE DEVELOPMENT

'If you think it's expensive to hire a professional, wait until you hire an amateur.'
– Red Adair

'Whether you want to uncover the secrets of the universe, or you just want to pursue a career in the 21st century, basic computer programming is an essential skill to learn.'
– Stephen Hawking

The aim of this chapter is to consider some of the ways in which people get started in the software development industry. With so many potential entry points, it is never too late to think about a role in software development, regardless of your background.

ENTRY POINTS INTO A SOFTWARE DEVELOPMENT ROLE

Malcolm Gladwell is a well-known journalist, author and public speaker. One of his most-quoted books, *Outliers: The Story of Success* (2008), includes a reference to what he calls the '10,000-hour rule'.

Simply put, this rule suggests that achieving success in any field requires 10,000 hours of practising related tasks. That works out at about 20 hours of work a week for 10 years. Only the core working hours are included – the times when your skills are focused and being ever more finely tuned by experiential learning.

Does this represent a true, aspirational metric for successful software development practice? Perhaps.

The IT sector is large, somewhat diverse and constantly growing, and demand for software developers is increasing as content requirements soar, particularly now that everyone has a supercomputer in their pocket disguised as a mobile phone.

The first generation of programmers, who learned programming using primitive 8-bit microcomputers back in the 1970s, is now heading towards well-earned retirement and, as you've probably heard, nature abhors a vacuum. Of course, software development project teams also hate empty developer chairs. This is where you come in!

There are multiple entry points into a software development role, and as such there is no guaranteed 'this always works' career plan. However, there are obvious things you can do to improve your appeal.

Getting started: age is no barrier

The first thing to understand is that, just like in any other occupation, developers come in all shapes and sizes. As such, entry points can occur at different ages. For example, junior developers may begin their careers at the age of 16, 18, 20, 30 or even beyond, depending on their life choices, educational background and relevant work history.

The following information will give you some ideas on how to begin your career in software development, depending on your situation.

Under 18s

If you are aged under 18, you are entitled to a free education, which can be studied full time in sixth form or at a further education (FE) college. You have the opportunity to study a combination of A-levels, T-levels and vocational options, all of which can be studied at Level 3 and can lead you into employment and/or higher education.

A-Levels

Most of the larger awarding organisations offer A-levels in computer science or computer science and IT. These focus on mathematical concepts, communication technologies, computer hardware and architecture, software, security, data modelling, algorithms, problem-solving, computer programming and file handling.

While this option might seem attractive because the content is broad, it is unlikely that you would be able to go directly into a job from an A-level programme because A-levels are classified as academic and are more theoretical than practical. Additionally, modern A-levels focus on examined assessment at the end of the course. With many development jobs now requiring candidates to undertake a practical programming task as part of the recruitment process, the lack of extensive practical coding in an A-level programme may make successfully completing the task a significant challenge.

If you are considering pursuing this option with a view to university progression, you should also consider studying mathematics at AS or A-level, as some higher education programmes now require this in addition to the A-level, especially for degree courses such as programming or software development.

Vocational qualifications

If you enjoy computing but prefer to be assessed with more coursework and fewer exams, the alternative to an A-level programme is to study vocational qualifications from a range of awarding organisations.

Pearson BTECs, City & Guilds and OCR Cambridge Technicals programmes have been developed in line with the National Occupational Standards for software developers (NOS n.d.). This means that their content is more directly targeted to the job with a much reduced focus on architecture, hardware and so on. Instead, there is much more emphasis on the software development lifecycle phases, including understanding the problem, designing a solution, programming, and then testing, implementing and maintaining a product. In this instance, a single course developed to meet these standards is equivalent to two or three A-levels and it will prepare you for either the workplace or progression into higher education.

Both A-levels and vocational qualifications are publicly funded for under 18s, which means that you can study without any cost to you or any requirement to repay the fees. In some circumstances, these courses can also be studied in FE colleges by students over 18, funded by a student loan, but this must be repaid when you are employed.

Apprenticeships

Apprenticeships are available to anybody over the age of 16. Level 3 is on the same stratum as an A-level, and in some contexts you can then progress to a Level 4 apprenticeship after completing your full-time studies at Level 3.

Apprenticeships are studied while working for an employer, often on day release, which means one day studying at a university or FE college or with a training provider, and four days at work learning on the job. The apprenticeship may well be a Trailblazer Apprenticeship. These apprenticeships were developed in partnership with employers to replace a range of outdated apprenticeships with a new format with more rigour, focused learning and a single End Point Assessment.

Because of the popularity of apprenticeships as an alternative to university, the government has created a website that serves as a central point for those looking for an apprenticeship opportunity: https://www.apprenticeships.gov.uk.

Whereas in higher education you have to cover the whole cost of the fees yourself (usually through a student loan), in an apprenticeship your employer will make a contribution to your education. You may, however, be expected to contribute to your own development (e.g. by attending courses and seminars in your own time, or even buying textbooks).

The National Apprenticeship Service has released a comprehensive list of all IT-related apprenticeships included under the heading 'Digital': see https://tinyurl.com/y3c95uaw.

Let's examine the different levels of apprenticeship available in more detail.

Level 2 (intermediate)

The entry point for apprenticeships is Level 2, which is usually referred to as 'intermediate'. This level is judged to be equivalent to five GCSEs at grades 4 to 9 (previously C to A*), with grade 4 considered a standard pass.

The current qualifications are still relatively generalist and cover a large area of study, as many of the fundamentals need to be learned before moving on to a Level 3 apprenticeship.

Level 3 (advanced)

Level 3 is equivalent to three A-levels at A* to E or a vocational Extended Diploma at pass, merit, distinction or distinction*. There are currently a number of IT apprenticeships that can be studied, including Cyber Security Technologist, Data Analyst, Digital and Technology Solution Specialist, Infrastructure Technician, Network Engineer, Unified Communications Technician and the Level 3 most suited to those wishing to pursue software development: Software Development Technician.

Successful completion of the Software Development Technician qualification prepares learners for junior roles in various fields, including:

- software development;
- mobile application development;
- web development;
- games development;
- application development.

In addition, there may be opportunities as junior or assistant programmers or automated test developers.

The Software Development Technician apprenticeship requires learners to focus on three key areas:

- knowledge and understanding;
- competence;
- underpinning skills, attitudes and behaviours.

The expectation is that learners will gain technical knowledge in areas such as security, data, analysis, quality and problem-solving, but also develop the ability to apply knowledge in an organisational or business context, together with an appreciation of the attitudes and behaviours that employers require. Employers expect their employees to be able to problem-solve, act professionally, work independently and, importantly, show that they can use their initiative. In addition, good communication skills are highly sought-after when you consider that a developer must be able to communicate effectively (and professionally) with clients and users, as well as their managers and fellow team members. The ability to work as part of an effective team is essential as very few development opportunities will ever be undertaken by an individual working alone.

The entry requirement for most Level 3 IT and digital qualifications is five GCSEs at levels 5 to 9, including English and mathematics.

T-levels

In addition to A-levels, vocational levels and apprenticeships, there are T-levels. These are designed to become the technical equivalent of A-levels, and the first T-levels will be available in schools and colleges from September 2020. In the first instance, these qualifications will not be available in every institution and a number of organisations across the UK have been selected to pilot the new schemes.

Based on the same standards as apprenticeships, T-levels will contain both a classroom element and a work experience or industrial placement. However, the key difference between apprenticeships and T-levels will be the amount of classroom study. In an apprenticeship, candidates spend most of their time in the workplace, undertaking less classroom study. In contrast, T-level candidates will spend most of their time in the classroom while also undertaking significant work-based activity (approximately 315 hours).

The first T-level for IT will be called Digital Production, Design and Development.

Level 4 (higher)

A Level 4 apprenticeship (also called a higher apprenticeship) can be attractive to those who wish to continue their studies but who do not want to access full-time higher education.

At this level there are significantly more apprenticeship options, but the two that are important in a software development context are Software Developer and Software Tester. This is not to suggest that software developers do not carry out testing activities (they most certainly do) – software developers carry out formative testing, eradicating errors as code is developed, and they also carry out summative testing at predefined points in development. However, there are new roles in industry for those who wish to focus on the testing phase of the development lifecycle in a summative sense, using a wide range of testing tools and techniques to improve test development outcomes before implementation and hand-over to the client.

The entry requirement for Level 4 IT and digital apprenticeships is three A-levels, a vocational Extended Diploma or a Level 3 apprenticeship, but again there is an emphasis on English and mathematics if these have not been achieved already.

After a Level 4 apprenticeship, candidates can progress to Levels 5, 6 and 7, where their technical learning is further advanced and combined with management.

Employer role and commitment

A relatively recent development in the world of apprenticeships is that there is now a greater role for the employer, who makes a financial contribution to the apprenticeship in addition to allowing the apprentice time to study. This makes the modern apprenticeship much more of a partnership between the employer and employee, both of whom make financial and practical commitments to the programme. You should be aware, however, that in some cases the employer may seek to recover some of their financial input from the apprentice if they immediately leave their employment at the end of the apprenticeship to go to another job.

Higher education

With a combination of Higher National Certificate (HNC), Higher National Diploma (HND) and degree programmes available across the UK, there are many potential routes through the higher education landscape.

A modern HNC is the equivalent of the first year of a degree, the HND is equivalent to the second year and the final top-up year of study makes up the full degree. HNC and

HND programmes can be studied at university or an FE college, although for the degree top-up you would probably need to go to university.

Each level can be studied either full time or part time, so you may be able to pursue these qualifications while you continue to work in your current role. You may even find that your employer will pay for, or make a contribution towards, your studies, particularly if they will be of direct benefit to the employer by making you more skilled.

Most higher education courses require you to take a student loan, which you will be expected to repay when you are employed.

Professional and vendor qualifications

Professional and vendor qualifications are also available, from a range of organisations, such as BCS, The Chartered Institute for IT, Microsoft, Oracle and Red Hat.

In recent years, professional bodies that represent the sector as a whole have developed qualifications with the heavy involvement of employer partners. BCS, for example, offers the Practitioner Certificate in Systems Development Essentials, which includes, on successful completion, a level of BCS membership. Professional qualifications such as this are not linked to a particular product but deliver a wide curriculum to candidates who already have technical skills and expertise but who lack the ability to apply their existing knowledge in a wider context.

A vendor qualification is one that is linked to a particular software or hardware component. Vendor qualifications have existed for many years, with Microsoft and Cisco being early adopters of the concept. Their vendor qualifications were specifically linked to their products and these certifications became valuable currency for candidates applying for roles in those areas. Chapter 17 examines these types of qualification in more detail.

You will study with a training provider or possibly at an FE college. Some can be studied remotely if there are online options.

Professional certification can be a very useful way of making you stand out in comparison to other job candidates. Typically, the first vendor or professional qualification has no specific entry requirements and is equally accessible to those with qualifications and those who do not have qualifications but have extensive industrial experience. Therefore, these qualifications can be an opportunity to get professional recognition for something you already know!

These courses are not usually publicly funded, which means that you will have to pay for them yourself or, in some circumstances, your employer may be prepared to pay. These courses are usually shorter – anything from a few weeks to a number of months – and are unlikely to require you to study for a full year.

With the cost of higher education still making qualifications at Level 4 and above inaccessible for some, apprenticeships and professional qualifications have seen a large surge in learner numbers in recent years. These qualifications have industry recognition and are increasingly popular for those who do not wish to face large student debts.

Already in the workplace?

Progressing from junior software development to a more senior role (such as senior or lead developer) can take considerable time and patience. Often, employers assess eligibility for promotion based on involvement in and delivery of successful projects. However, there are some simple steps to follow when opportunities arise:

- If you are offered training in new technologies or techniques, **always** do so, especially if it is an industry-recognised certification.
- If you are given the opportunity to stretch into new areas, generally it is a good idea to accept. Any learning, especially that which is conducted 'on the job', can be very rewarding.
- Pair programming techniques (see Chapter 7) are beneficial, particularly when you can work alongside a more experienced developer.
- Keep your programming skills up to date, and embrace new concepts and ideas. Working on a pet project outside work hours is a fun way to hone your skills without extra pressure.
- Grow with new responsibilities – particularly in terms of leading teams and undertaking different roles.
- Above all, enjoy challenges and solve them as best you can!

SOFTWARE DEVELOPER SKILLS

In the context of software development, across the whole lifecycle, the skills needed can generally be divided into two types – transferable skills and development skills. These include (but are not limited to) the skills below.

Transferable skills

These are skills which you develop during your working life that can be transferred and used in different contexts.

- **Ability to work individually and as part of a team:** most software development is based around team activity unless the project is small enough to be undertaken by an individual, but this is increasingly unlikely given the relative complexity (and technological diversity) of modern projects.
- **Active listening skills:** this involves listening to the opinions of your team and the needs of your client, and concentrating on what is being communicated and the way it is being communicated, as non-verbal communication often adds important context.
- **Emotional intelligence:** this means being co-operative and compassionate when dealing with peers, and being constructive and supportive in your contributions, especially when giving feedback to others.

- **Negotiation skills:** you must be able to have discussions and, more importantly, make compromises to agree a resolution that suits both parties. This is not always as easy as it may sound, particularly when you are trying to balance the needs of a large group of stakeholders and a range of business drivers.
- **Written and verbal communication skills:** you must be clear and (preferably) concise both in writing and in discussions and presentations to your client. To achieve this, you must ensure that you are fully prepared by gaining a comprehensive understanding of the business needs, known constraints and technical challenges that lie ahead. Periodically immersing yourself in a new business sector is usually also a good idea.
- **Attention to detail:** ensuring documents, emails, discussions and so on are complete and correct will help to make sure that every member of the team (and the client) has a full understanding of the development process and current progress, targets and goals. Missing out key details can have catastrophic effects in a development process.
- **Problem-solving:** not everyone is a natural problem-solver but most of us have valuable insights and contributions we can make as part of a larger design or implementation process. The key is knowing when these are appropriate (or not) and how to effectively express your ideas. Formalised problem-solving provides the developer with a guaranteed step-by-step approach: (1) understanding the problem, (2) planning a solution (considering alternatives) by breaking it down into smaller problems and (3) finally implementing the chosen solution. Everything else (languages, frameworks, libraries etc.) is just the **tools** you use to support this activity and these will undoubtedly change over time.
- **Project management:** managing software development relies on two key components: effective and realistic planning, and appropriate (thorough) monitoring of activities. A good project manager will ensure that a product is delivered on time and within budget.
- **Trouble-shooting:** unlike problem-solving, which is usually carried out in the context of the development and its deliverables, trouble-shooting requires much faster responses, immediate reaction to issues and an ability to find resolutions quickly.
- **Mentoring skills:** everyone has to start somewhere, so less experienced team members should be supported to help them to build confidence that they are approaching the development in the right way. This will make them better team members for future developments.
- **Good understanding of business context:** this is the most difficult of the transferable skills to acquire and is almost always only achieved with experience. There are, however, some key areas you can investigate to give you at least some theoretical understanding of business context:
  - be able to identify the different functional areas of an organisation;
  - understand what each functional area does;
  - understand what data the functional areas need to operate and what each generates as a result of its activities.

Development-specific skills

Some skills are traditionally associated with the role of a software developer. These include but are not limited to:

- **Analysis skills:** good analysis is based on a person's ability to examine something in detail. Superficial analysis will almost always become a factor that leads to the failure of a software development project. Sometimes analysis is less in-depth due to time constraints, which is why your approach should be planned to ensure that the analysis you undertake is focused and is likely to produce the level of information that you need. Analysis may be needed in software development for various reasons – for example:
    - determine how to access data items from a complex data structure;
    - debug complex algorithms to improve their efficiency;
    - assess benchmarks from code execution to determine which techniques are fastest;
    - make development choices based on research into good coding practice.
- **Programming (language-specific syntax but also a wider understanding of programming constructs, tools and techniques):** experience using the language that will be used in a development project is key. Some would argue that understanding common algorithms is a transferrable skill and that all that differs is the actual language syntax. While that may be true to some degree, you will find differences even between versions of the same language which can slow you down when coding (e.g. changes in the way components work or where you find them within the environment). You must make sure that you are fully oriented within the relevant language and, if it is completely new to you, ensure that you take time to learn key techniques and that you know how to access essential tools. It is also worth considering the evolution of programming languages; for example, coding in **classic** C++ is a very different proposition to doing so in **modern** C++.
- **Structured Query Language (SQL) skills:** as shown in Chapter 9, most applications use database connections as a source of external data, and as such you should be able to perform the basic data operations using the world's most popular database language, for example:
    - SQL Data Definition Language (DDL), which is used for creating databases and tables, altering structures, renaming, dropping (removing the table and its data) and truncating (deleting all data from a table);
    - SQL Data Manipulation Language (DML), which is used for inserting, deleting and updating rows;
    - SQL Data Control Language (DCL), which is used for granting and revoking user privileges to database objects;
    - SQL Data Query Language (DQL), which is used for creating queries using the `select` statement (probably the most important).
- **Hardware skills:** it is always a good idea to have a basic working knowledge of the technology used by your applications – for example, you should understand

how programs use computer memory to enable you to avoid memory errors caused by poor code. Another example is understanding the architecture of the central processing unit (CPU) so that code can be more efficiently written and take advantage of the hardware's natural features.

- **Networking skills:** many applications will be network aware, whether these are traditional desktop applications, mobile applications or commercial web applications. As such, it is useful to have a working knowledge of basic networking principles such as addresses, protocols and sockets.
- **Knowledge of an operating system (OS):** software developers' role is not limited to producing program code; they also have to prepare development environments and deployment environments (i.e. where the code will eventually be run). This very often involves installing software, managing configuration files and working with files (and permissions). Much of this should be done from the operating system's command line interface (CLI) – that is, Windows' command prompt or Linux's shell – rather than the graphical user interface (GUI), although note that on a server the latter may not be available. As such, a working knowledge of the target operating system is crucial.
- **Regular expressions:** often abbreviated to 'regex', this is a popular tool used by experienced developers and supported in most modern programming languages and applications. This intricate and elegant pattern-matching metalanguage can be used to match, extract or substitute data in a string without requiring the use of complex and messy combinations of different string functions.
- **Diagramming skills:** a diagram is often much more useful at explaining existing or new systems than extensive text. There is a host of software applications that you can use, such as Lucidchart, SmartDraw and Draw.io. They are fundamentally very similar in that you work with shapes, arrows and lines to produce a representation of a system or processes within a system.

Flowcharts example

The flowchart in Figure 1.1 was created using Lucidchart. The flowchart represents the simple process of issuing a bus pass to travellers aged 60 or above.

Firstly, the user inputs the age of the traveller. The program then evaluates the age and either issues the bus pass (if the age is correct) or ends the program.

If you investigate different diagramming software you may see variations in the shapes. It is important that whichever version you use, you use it consistently.

You might find it helpful to identify a particular software application and practise using it to generate diagrams of systems and processes.

Figure 1.1 A simple flowchart

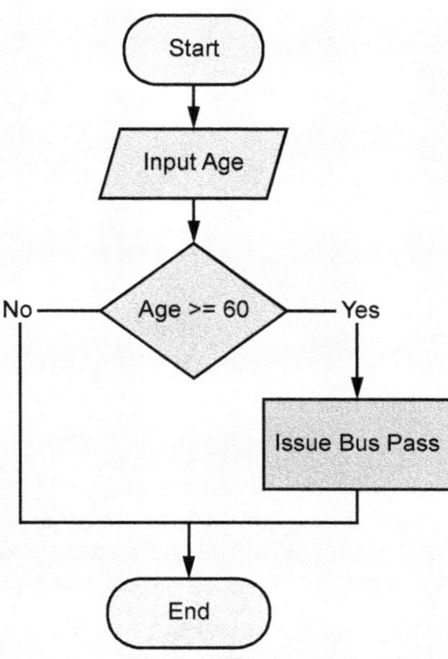

- **Design skills:** once you understand the problem to be addressed and the requirements, you may well be able to visualise a solution or components that will contribute towards a solution. Designing a solution as part of a team is often a good way of producing a high-quality product as you can draw on a large range of ideas.
- **Technical writing skills:** these skills enable you to produce a range of important documentation, including:
  - development documentation, which begins with a copy of the development brief and then records details of the investigation, the specification, the design, the implementation and testing activities, and (particularly importantly) the rationale for the decisions made;
  - technical documents that explain the final system as developed to aid future maintenance and upgrades;
  - user documentation to support new users of the software.

Tips for user guides

- Avoid waffle. Users appreciate short, succinct and accurate instructions to help them navigate the software.
- Ensure that you consider users with disabilities by offering the guide in different formats.
- Ideally, test the guide on users during the user-test part of the development process.
- Documentation may be paper-based, electronic or a combination of both.

- **Estimation/costing skills:** these are essential if you are required to prepare a costing or estimate for a client. Keep in mind the following:
  - Ensure that you understand the objectives of the development.
  - You must have a full understanding of the scope of the project (this will ensure that your project does not suffer from scope creep). Scope creep is where the project expands into unexpected areas which were not part of the agreed development (see Chapter 4 for more information on scope creep).
  - Consider what will be needed in terms of both functional requirements (what the software is expected to do) and non-functional requirements (which will also include usability, security, and interoperability with other systems and maintenance).
  - Consider adding additional time to facilitate recovery points for when the development project experiences problems.
  - Consider using project management estimation tools, either to help you to carry out the costing or to confirm the results of a costing you have undertaken by other means.
  - Your skills in this area will improve over time; however, as you are building up experience, draw on the knowledge and experiences of others in your team.
- **Testing:** programs which have not been robustly tested can be disastrous for a number of reasons. The program may fail under load (large number of users) and may become slow. Alternatively, functionality may not work as expected. Testing is about planning, executing, documenting and resolving issues to make sure that the product is fully functioning and meets the needs of the client and the users. This requires a highly structured approach to ensure that the testing is appropriately comprehensive. As a developer you will choose the range of data and the data items that will be used to test the software product. A popular technique is the use of unit testing, which devolves testing responsibility to small parts of the solution (e.g. individual functions which perform a specific task). More on testing can be found in Chapter 10.

INTERVIEW SKILLS

Seeing a suitable vacancy and applying for it are only the first steps. Once you have received an invitation to attend an interview, you need to prepare in advance. Simply turning up and hoping for the best will probably not result in success.

Firstly, don't panic. You will be there on merit, having impressed the interview panel through your CV, covering letter and any other documentation you were required to send. While it is outside the remit of this book to describe good interview technique, we can pull back the curtain to discuss common testing strategies you might face in an interview.

Within the IT sector, it is quite common for software development applicants to sit at least two timed tests:

- **An intelligence test:** this is often a written test which combines elements such as numerical reasoning, letter sequences, vocabulary, spelling, punctuation, lateral thinking and homonyms (two words with the same spelling but different meanings, e.g. 'bright').
- **An aptitude and ability test:** this will be designed specifically for software developers and typically involves solving a programming problem. Often it isn't very complex – for example, you might be asked to create graduated bonus features for a simple guessing game, such as personalisation and a table of high scores. What is significant is that the time limit will be deliberately tight, forcing you to make quick decisions and prioritise your problem-solving. It is quite common for the interview panel to provide constructive feedback on your efforts, offering you a chance to explain your design and development decisions. This can be critical as the panel will usually be genuinely interested not just in your coding abilities but also in your problem-solving and decision-making, particularly when under pressure.

An alternative test might involve completing programming tasks using a remote learning platform which times and grades your efforts. Ultimately, it is up to the organisation how it chooses to judge your ability and performance. For more senior roles, a screening telephone interview may be used initially.

In short, any testing is designed to do two things: put you under pressure and see how you solve problems. Both are completely representative of the everyday challenges you will face as a software developer.

TIPS FOR GETTING STARTED AS A SOFTWARE DEVELOPER

The following tips will help you to get started as a software developer:

- Be 100% committed to this career pathway.
- Target an initial programming language and/or technology stack to learn – it's generally not too difficult to transfer your skills to a new one.

- Program for fun, not just work. Personal projects are rewarding and permit time for experimentation without the risk of impacting ongoing professional projects.
- Use industry tools and workflows – for example, use popular version control tools and gain experience with different command line environments, current project management platforms and so on.
- Don't insulate yourself; investigate wider technologies that often dovetail with yours (e.g. CPUs and system architecture, networking, web standards, databases and emerging cloud technologies). This will help you to communicate effectively with stakeholders in other disciplines.
- Learn to read technical documentation. Much of a software developer's job is working with new libraries, web-based application programming interfaces (APIs), operating systems and so on. You may need to delve into written material to find answers on those rare occasions when a search engine can't help.
- Be prepared to work under pressure and within strict time limits. From a job interview's programming test to partaking in weekly sprints, the challenge of getting the job done within the given constraints to the best standard possible is ever present.
- Be aware of potential vulnerabilities and the best defensive coding practices in order to reduce the likelihood of successful cyberattacks and improve the robustness of your solutions.
- Don't have a precious ego. Everyone, no matter how good they are, can learn from other software developers (even relatively inexperienced ones). Read other people's code to absorb new ideas and accept feedback and criticism of your code without taking personal offence – everything can (and should) be improved over time.
- Be prepared for rejection. The industry is highly competitive, and you will undoubtedly end up vying for a job with other developers who have more experience.
- Invest (time and money) in your skills. Regularly attend workshops, seminars, training courses and so on.
- Be active in the software development community – whether you publish a blog, have a LinkedIn profile, lead an open source passion project, contribute magazine articles, or attend conferences and meetings, be visible. It will help potential employers to form a positive impression of you.
- Be prepared for lifelong learning. IT is a fast-changing sector to work in, and software development doubly so. Your skills need to evolve to meet the challenges and demands that will be placed on you.

SUMMARY

As you can see, there is no single route, educational background or area of life experience that is mandated for stepping into the role of software developer. Gaining recognised qualifications will almost certainly be helpful and many different options exist. Training in the workplace is also invaluable and can be combined with vocational qualifications to good effect.

However, don't think you'll stop learning once you've got your desired role! The IT industry, and most notably those who develop professionally within it, face a constant battle to keep their skills up to date and competitive.

The next chapter examines target roles in the software development industry.

2 TARGET ROLES

'Software developers design, build and test computer programs for business, education and leisure services.'

– UK National Careers Service (2020)

There are many ways of differentiating software developer roles. This chapter considers the roles that exist and how they are conceptualised by different bodies (e.g. government and recruitment). It also outlines a potential route through the recruitment process.

OVERVIEW OF DIFFERENT ROLES

The Level 4 Apprenticeship Standards (as contributed to by BCS) provide a short overview of the tasks of a software developer: 'building and testing simple, high-quality code for software' (Institute for Apprenticeships & Technical Education 2020). They identify five typical job roles:

- **Web developer:** typically front-end, back-end or full-stack.
  - Front-end: focusing on commercial website UIs (user interfaces), for example using HTML, CSS and JavaScript (vanilla and frameworks).
  - Back-end: focusing on server-side scripting, for example using PHP, Microsoft C# (.NET), Oracle Java, Python, Node.JS and external data sources (such as relational database management systems).
  - Full-stack: focusing on the complete technology stack, which means everything from the back-end through to the front-end. This might also include 'enterprise glue' (e.g. Bash script, Perl or PowerShell) for linking back-end systems together and any other manner of supporting technologies.
- **Mobile app developer:** developing native applications often specific to a certain mobile operating system (e.g. Apple iOS or Google Android OS), although hybrid applications are possible.
- **Games developer:** designing interactive experiences in the PC, online, mobile or games console sectors.
- **Software developer:** a more generic catch-all descriptor which is often used by a smaller employer with specific needs that don't neatly fit just one of the above role types.
- **Application developer:** another generic descriptor (see above).

Although these roles appear to be different, as you would expect they all depend on certain common skills, such as analysis, design, implementation and testing. However, in contrast, they all rely on different programming languages. For example, a mobile app developer might focus on Java (for Google Android OS) or Swift (for Apple iOS) whereas a web developer with a front-end remit is unlikely to rely on these.

GOVERNMENT PERSPECTIVE

From the UK National Career Service's perspective, the role of a software developer sits alongside related roles such as solutions architect, web developer, computer games developer, IT project manager and robotics engineer.

For further information see https://nationalcareers.service.gov.uk/job-categories/computing-technology-and-digital.

The Digital, Data and Technology Professions (DDaT) framework

The UK government has a Digital, Data and Technology (DDaT) Capability Framework which specifies the roles and skills required for each identified job in the Civil Service. There are currently six 'families' in this profession:

- data job family;
- IT operations job family;
- product and delivery job family;
- quality assurance testing (QAT) job family;
- technical job family;
- user-centred design job family.

The role of a software developer fits within the technical job family, and the following duties are identified:

- 'be responsible for writing clean, secure code following a test-driven approach';
- 'create code that is open by default and easy for others to reuse' (Gov.uk 2020).

This role is broken down into several sub-roles (listed in order of seniority):

- apprentice developer;
- junior developer;
- developer;
- senior developer;
- lead developer;
- principal developer.

In the three most senior sub-roles, it is anticipated that individuals will have both technical and managerial responsibilities – for example, understanding security (technical) and guiding others in modern standards (managerial). In practice there are often discrete roles allowing developers to progress and align their careers more to one area or the other – for example, lead developer (mostly technical) versus project lead (mostly managerial).

In all cases, the responsibilities associated with each sub-role are classified with a relevant skill level:

- **awareness:** know the skill and how it's applied;
- **working:** have knowledge and experience, and adopt the techniques that are most appropriate;
- **practitioner:** share knowledge and experience with others, defining the techniques that are most appropriate;
- **expert:** recognised as a specialist and adviser in the skill, leading or guiding in its best practice.

Putting this in context, here are some examples of sub-roles, their responsibilities and their associated skill levels:

- **junior developer:** has a working knowledge of information security;
- **developer:** has a practitioner's ability to prototype designs and iterate them over time;
- **lead developer:** has an expert ability to advise on the best way to apply standards and methods when building solutions, particularly in the use of programming tools and techniques.

As you can see, these descriptors are quite nuanced and, perhaps, can give you a real feel for the progress ladder associated with the role of software developer. They are certainly worth a read (see Gov.uk 2020).

RECRUITMENT PERSPECTIVE AND SENIORITY

Job roles offered by recruitment agencies often specify a combination of role type **and** seniority. It is also quite common for recruiters to include a specific technology or programming language as this helps to filter applicants much more quickly and can be used to pattern-match against applicants' CVs on file.

Common examples of advertised job titles include:

- junior web developer (HTML, CSS, JavaScript);
- senior full-stack developer;
- experienced back-end PHP developer;
- team leader for Android mobile app development.

The timeline for software development recruitment might follow this common flow of events:

1. The candidate will register on a recruitment website and upload their CV.
2. The recruiter will select potential candidates based on CV match algorithms (role, location, experience, skills etc.).
3. The candidate will filter recruiter emails, identify suitable opportunities (based on role, seniority, salary, skill etc.) and contact the recruiter to express an interest.
4. The recruiter will forward the candidate's CV to the employer and then (if the candidate is shortlisted) arrange an interview with the employer.
5. The candidate will complete a telephone screening interview (or online assessment), attend a face-to-face interview and/or practical assessment, provide references and (if successful) accept an offer of employment.

Of course, it's not an exact science and it may take multiple attempts to land the role you really want; don't get discouraged! It is always worth remembering that some roles, particularly that of junior developer, are keenly sought-after and therefore often over-subscribed – after all, it's the first step on the software developer's career ladder.

TIPS FOR TARGETING ROLES

If you are unsure, research the roles and sub-roles so that you can:

- understand the different sub-roles that are available and their responsibilities;
- develop the practical skills and knowledge required by each sub-role;
- use experience to tune these skills to contribute to successful projects;
- curate a digital portfolio which showcases your skills (this is easiest for web or mobile developers but publicly hosted open source projects are useful assets too);
- be prepared to prove your technical skills under testing conditions (and a time limit!).

Above all, be patient – don't be afraid of rejection, ask for feedback and use it to improve your next application.

SUMMARY

In this chapter we have reviewed job roles associated with software development and examined a typical progression ladder in the sector.

In the next chapter we will examine the different tasks associated with software developer roles.

3 OVERVIEW OF DIFFERENT TASKS A COMMERCIAL DEVELOPER MIGHT ENCOUNTER IN THE ROLE

'What's in a name? That which we call a rose / By any other name would smell as sweet.'

– William Shakespeare

In this chapter we'll investigate how a commercial developer's day-to-day tasks vary depending on a variety of factors.

WHAT'S IN A NAME?

The first thing to think about when you become a software developer is that the term means many different things to many different people, and this can be doubly true of employers. Generally, the recruitment of junior and senior development posts is overseen by a lead developer, a project lead or manager, or even (in smaller organisations) a chief technology officer.

Although template descriptions can be found across the internet, there is no standard title or job description that you will find across job adverts or recruitment websites. So how might the role be differentiated? We could differentiate the role as follows:

- by **entry level** (as discussed in Chapter 2), that is:
    - junior developer;
    - experienced (mid-level) developer;
    - senior (lead) developer.
- by **programming language**, for example:
    - Java programmer;
    - C software engineer;
    - Python developer;
    - PHP coder.
- by **type of system**, for example:
    - commercial web developer;
    - mobile app programmer;
    - systems software engineer;
    - database programmer.

- by **specialised core responsibility**, for example:
  - system analyst;
  - system designer;
  - software/systems tester.

Any combination of these titles and criteria could be used in conjunction to describe a post, for example 'junior Python developer', 'Java mobile app programmer' or 'JavaScript commercial web developer'.

WHAT WOULD MY RESPONSIBILITIES BE AS A SOFTWARE DEVELOPER?

The titles of jobs often don't describe a role fully. For example, a commercial web developer's responsibilities may be quite diverse and, depending on the seniority of the role, encompass many different skills and technologies.

Another factor to consider in thinking about responsibilities is the size and type of organisation involved – responsibilities may vary greatly between similarly named roles in a large public sector organisation (such as the NHS) and a smaller private company (e.g. with fewer than 50–200 employees). It really can be quite confusing! But don't abandon all hope: the simplest place to start is to examine the typical essentials and desirables that an employer might include in a software developer's job description. Truth be told, this can be a bit of a project manager's 'bucket list', containing the most common responsibilities that a developer might be called upon to fulfil. It's unlikely that any candidate will be able to meet **all** the criteria to the same exacting standard; some you may need to grow into over time, for example by learning on the job.

However, a project manager can dream, so here is a typical list of software developer duties to consider:

- attend meetings with clients (including leading presentations);
- analyse client and end-user requirements;
- create technical specifications for development;
- write **new** program code (in multiple languages) to organisational standards;
- modify **existing** program code to improve its performance or efficiency, or add new desired functionality;
- perform system integrations through data exchange or enterprise glue techniques, including for legacy systems;
- write technical documentation;
- write user documentation;
- test the functionality of program code and identify and rectify deficiencies;
- work as part of a team consisting of many different related disciplines, for example UX (User experience) designers, graphic artists, other developers, system testers and project managers;

- investigate new technologies and techniques;
- install, update or configure new software libraries, services, systems or hardware as required;
- partake in continuing professional development by updating existing knowledge and learning new techniques;
- observe the principles of the GDPR (General Data Protection Regulation).

You may think that's a long list, but it's genuinely not that unusual, particularly in smaller organisations. Often it's the case that the smaller the company, the more of the listed duties a developer may need to perform.

SUMMARY

In this chapter we have explored the types of activities you may encounter in the role of a software developer.

The next chapter examines the different methodologies used to develop software.

4 OVERVIEW OF SOFTWARE DEVELOPMENT METHODOLOGIES

'Things change. The only thing constant is change. It's up to you to be adaptable.'
– Anonymous

Anyone who tries to make a cake simply by guessing the recipe and method will very quickly discover that the outcome may not be a good cake. Most of us will reach for a recipe book. In the same way, when embarking on a software development project, using a methodology to provide a framework and sense of direction will ensure a better outcome. This chapter explores a range of possible approaches.

DEVELOPMENTAL APPROACHES

At the heart of a new software development project is the need to solve an identified problem. But what is the best way to tackle this and communicate ideas?

Unfortunately, there isn't a single 'one size fits all' framework, methodology or technique that will solve all your problems – that would be too easy! Moreover, different organisations adopt different methods and tactics to solve their problems. However, there is a reasonable starting point and that is the software development lifecycle (SDLC).

The software development lifecycle

A software development project usually follows a set process with a series of defined (and predictable) steps. This is known as the SDLC. It usually begins with the identification of an internal problem or aspiration, often in the form of a brief. Reviewing the product at the end of the lifecycle can trigger a new development.

The lifecycle contains a series of phases, each one of which contributes a process to the development (see Figure 4.1).

The length of a SDLC varies greatly depending on the size and complexity of the problem it is attempting to solve. Despite this, the lengths of each stage generally remain proportionate – for example, the coding phase is longer (potentially many months) compared to a shorter design and architecting period (perhaps several weeks).

The basic components of the lifecycle (up to the point of roll-out) are as follows.

Brief

This triggers the development process. It could be based on a problem with an existing system or process, something new that is necessary to meet a business requirement, or

Figure 4.1 The traditional phases of the software development lifecycle

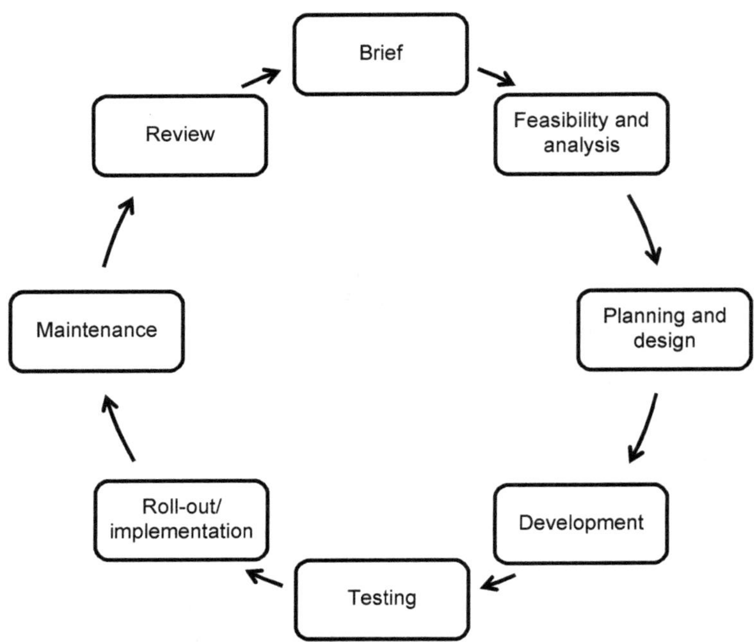

simply a response to an emerging opportunity based on a new idea. It may occur as the result of an organisation's aspiration to provide a better customer experience, improve its processes or reduce costs (or all three!).

Feasibility

This is a top-level investigation into the brief which essentially considers the problem to be solved and investigates potential solutions. These solutions can include:

- **An off-the-shelf solution:** an existing product or application that can be purchased and used without modification.
- **A tailored solution:** a generic application is used to build an appropriate solution (such as a spreadsheet or a database package), or an off-the-shelf solution is adapted with additional functionality to meet the organisation's needs.
- **A bespoke solution:** a solution that is written from scratch with a full process of investigation, design, development and testing before implementation.
- **No change:** there are times when the costs of a development project would far outweigh the organisational benefits that would be achieved. This is often measured financially in terms of the potential return on investment (ROI). ROI is calculated to evaluate how well a particular investment will perform (or has performed). A low ROI would suggest that the software development might not in fact be feasible.

See 'How the Client Brief Affects the Development Process' later in this chapter for more on potential solutions.

Analysis

Once it has been established that a development is required, a full analysis of the problem and the existing system (and how it could contribute to the solution) is undertaken. This is documented to ensure that subsequent decisions can be justified.

Planning

This is a very necessary, formal part of the process which can sometimes be overlooked. It is where expected activities and processes needed to achieve the solution are identified, along with anticipated timeframes. Available resources are then established and allocated to the tasks that will need to be undertaken.

Design

The completion of the analysis leads to the design of the new system. This may or may not be formally documented (depending on the nature of the development and the methodology being used to focus activities).

Development

This phase involves the actual coding of the new solution, system or service.

Testing

A range of testing strategies may be applied in software development. Some are undertaken during development, others when the development is considered relatively complete, and others during a period of final testing before the developed system is handed over to the client.

Roll-out/implementation

Once the code has passed testing, it will work flawlessly in a staging environment and be ready to go 'live' and into production. At this point, the system is deployed (installed and prepared) for use by internal and/or external customers.

The client will look to the developers for advice about the implementation of the software or system so that it can be integrated appropriately into the business (often called a 'changeover strategy'). There is no single right answer to the question of how the solution should be implemented, which is why the development team must consider the needs of the client, the risk factors, and the amount of additional work that will be incurred through the implementation activity in the advice they give.

Common to every changeover strategy is the need for the organisation's data to be added to the new system so that when it goes live it has something to work with. This may require changes to the data to prepare it for integration, such as converting data formats or reorganising the data to prepare it for transfer. The data itself will usually be moved to the new system at a time when normal business transactions are suspended because the organisation is closed. This can be more challenging if the business operates online.

Table 4.1 explores the most commonly used changeover strategies, the risk factors and additional considerations related to them.

Table 4.1 Changeover strategies

Changeover strategy	Risk factors	Additional considerations
Direct: once data has been added, the existing system is switched off and the new system takes over.	• High risk • If the system fails, business could be interrupted, which could result in lost data, impacting revenue and/or reputation	This is the most straightforward changeover strategy because it does not require any additional work.
Parallel: the existing and new systems are run together. This allows the operation of the new system to be checked by comparing it to the output from the existing system.	• Low risk • Errors will be quickly identified • All transactions will be recorded on a system which is known to work correctly so it effectively acts as a backup	This is the most labour intensive changeover strategy because every transaction has to be recorded twice. It is only a short-term solution as it creates additional stresses and workload for staff.
Pilot: the new system is introduced in one area of the organisation first, such as a branch office or shop. When this has been proved successful, the system is rolled out to other branches or shops until all parts of the organisation are using the system.	• Reduced risk in comparison to a direct changeover • Errors will impact a smaller area of the organisation • All parts of the system can be trialled • Staff can cascade training to other areas as they come on board	The main issue is that as with a direct changeover, the existing system is switched off, so the risk factors are similar. However, the risk is less as any errors would impact a smaller area of the business.
Phased: this introduces the new software in steps or stages. As an example, consider the sales function of an organisation where the sales force is split into geographical teams. One team could be moved to the new system and, when this has been proved successful, another team could be added.	• Reduced risk in comparison to a direct changeover • Errors will impact a smaller area of the organisation • All parts of the system can be trialled by teams • Staff can cascade training to other areas as they come on board	The main issue is that as with a direct changeover, the existing system is switched off for those being moved across, so the risk factors are similar. However, the risk is less as any errors would impact a single team or a small number of teams in an area.

Considering a soft launch

A common tactic employed in the IT sector is the use of a so-called soft launch. Prevalent in software releases and commercial websites, in a soft launch the product is made available to a selected audience without any fanfare or marketing.

For a software product this can be used as a form of beta testing, allowing the organisation to test its product 'in the wild' without its reputation being affected if the results are not favourable. In addition, it will receive useful bug reports and usage statistics that will help it to resolve identified issues or further improve the product's performance before full release.

Commercial websites often use such soft launches to roll out new features or make modifications of the user interface. Again, user feedback is invaluable in shaping the next stage of development.

Maintenance

Designed. Coded. Tested. Implementation complete. The new solution is in production and the client appears to be happy. The next step is to ensure ongoing maintenance. Once development is complete and the system has been implemented, a maintenance strategy is put in place to support the system going forward.

Once it is operational, the system is examined for any issues that have arisen as a result of its use and identified errors are fixed. For example, a user might complain about tabbing errors, which is where input boxes are not presented to the user in the right order and the user finds themselves jumping around the screen. This does not indicate that the product is faulty; it is essentially an irritation that really should be eradicated but which probably will have a low priority. If, however, a user discovers that sales records are not being correctly committed to the sales database (for instance), this would be a high-priority issue that would need to be resolved quickly.

Maintenance activities usually fall into one of the following three categories.

Corrective maintenance

Corrective maintenance is reactive. It is usually prioritised because it aims to resolve significant problems with the system that have emerged during its use. Bugs and other defects need to be eradicated quickly for two reasons: firstly, because the product will not be working as it should until these are resolved, and secondly, because users may well view the product with increasing negativity if bugs persist.

Typical bugs include calculations not outputting accurate answers, and button clicks or hyperlinks not working.

Perfective maintenance

Perfective maintenance is also reactive. Following corrective maintenance, perfective maintenance resolves other issues. These issues do not prevent the system working but are known to be irritating to users.

Tabbing errors (as mentioned above) are one example of an issue that would be addressed during perfective maintenance. Another example could be the slow loading of a web application, caused by too many extraneous HTTP (Hypertext Transfer Protocol) requests during initial loading. In such cases a developer will use specialised tools to review and improve the design. For example, a pipeline profiler might be used to analyse the performance of a web page while it loads. This type of activity can often be the developmental equivalent of diving down a very dark and deep rabbit hole, and it may lead to the unearthing of graver concerns, such as CPU load issues and memory usage inefficiencies. Perfective maintenance is time-consuming but it may not garner the same amount of resources as corrective maintenance, which is perceived to be more critical; in other words, perfective maintenance is a 'nice to have' but not essential.

From a purist's point of view (and in an ideal world), anything that negatively impacts a user's experience of a software product should be considered critical. However, time and money are always at a premium in any development project, so higher-priority issues will typically win out.

Adaptive maintenance

This is a longer-term form of maintenance and is usually a response to the organisation's ever-changing environment or its future aspirations. New technologies could emerge which the business would like to exploit. The organisation may have expansion plans, such as selling new product lines or services which will mean higher transaction volumes or larger numbers of users. In some cases, adaptive maintenance may even be required in response to the activities of business rivals and the need to be more competitive.

Adaptive maintenance is proactive because the organisation is able to plan the maintenance activities that will be needed to prepare for its future plans.

It is certain that your organisation will be involved in corrective maintenance, and it is likely that it will also be involved in perfective maintenance. Adaptive maintenance may well initiate another development project and the SDLC will start again.

Review

At the end of the SDLC, a review is undertaken which not only examines the success (or failure) of the product but also examines the development process (from the receipt of the client brief through to the implementation of the software), decisions taken and so on. Usually involving a range of stakeholders, this activity provides the development team with an overview of their activities and what went well or badly. It also provides an opportunity to consider how future developments could be improved based on lessons learned.

A good project manager will produce a checklist that covers all of the development activities so that they can be individually inspected. This will provide a useful focus for the evaluation process and a framework for evaluation, enabling subsequent development activities to be examined in the same way. In this way, it can prove or disprove whether lessons learned have really affected the organisation's development processes.

Here is a very basic checklist of what could or should be considered under each phase:

1. Brief
    - How well was the client brief understood at the start of the development?

- How much additional information had to be sought?
- Did this change the nature of the client brief or reinforce what was originally known?
- Was a requirements specification formally produced and agreed between the client and the development team?
- Were constraints and limitations correctly identified?
- How well were the client's agreed instructions communicated to the development team?

2. Feasibility/analysis
   - Was a feasibility study undertaken?
   - Would a feasibility study have provided a wider range of potential solutions?
   - Was the analysis suitably executed?
   - Were appropriate investigative techniques used?
   - Was any information overlooked which would have benefited the development?
3. Planning/design
   - Was a plan produced?
   - Was it sufficiently detailed?
   - Was enough recovery time built in to accommodate problems experienced along the way?
   - Was the design comprehensive?
     - Did it cover all functional requirements?
     - Did it cover all non-functional requirements?
     - Did it sufficiently cover the user interface design?
   - Were the techniques used to articulate the design appropriate?
   - Were the coders able to interpret the requirements from the design documentation?
4. Development
   - Was enough time given to coding activity?
   - Was the right guidance given to individual coders or coding teams?
   - How well did the teams work together to produce the solution?
   - Was the right language used to produce the solution?
   - Did the coders have sufficient skills or was additional support needed?
5. Testing
   - Was the testing suitably planned?
   - Was sufficient time given to the testing activities?
   - Was appropriate test data generated for each testing phase?

- Were the outcomes of the testing analysed to produce a comprehensive plan for corrective action?
- Was sufficient user testing undertaken to be confident of future user acceptance?

6. Roll-out/implementation
    - Was the changeover strategy adopted right for the context?
    - Were the implementation risks correctly identified and was action taken to mitigate the identified risks?
    - What difficulties were experienced?
    - What could have been done differently to smooth the transition process?
7. Review
    - How should the product be supported and maintained going forward?
    - What will we as an organisation offer in terms of support?
    - Is there a particular process that should be set out to ensure that the product is effectively maintained?
    - Has the review considered the views of **all** stakeholders and stakeholder groups?
    - Did the differing needs of the groups impact on the success of the development project?
    - Did the development team have sufficient access to information to be able to identify what went well and what went badly?
    - How should future development activities be modified to ensure similar mistakes are not repeated?
8. General
    - Was the right methodology adopted to support the development?
    - Were any aspects of the methodology less than helpful in supporting the development activity?
    - If the project were repeated, would the same methodology be chosen?

The review phase of the development should be extensive and the activity should be undertaken with commitment, particularly as there may be a tendency to want to end the current development project so that the next one can begin. Understanding what went well and where improvements could have been made will be invaluable experience for the next development.

HOW DEVELOPMENTS GO WRONG

A suspicious but worldly wise project manager once said, 'What could possibly go wrong?' Given the long gestation time of many projects, the answer is likely to be, 'Quite a lot.' (For example, see Maughan 2010.)

Here are a few things to think about – always remember, forewarned is forearmed!

The perils of scope creep

We previously discussed the importance of formally agreeing the boundaries and requirements of a project at the outset to manage the client's expectations and to focus the activities of the development team.

In an effort to keep the client happy, it is very common for developers to try to accommodate requests from the client that emerge during development. Unfortunately, the client often has little or no understanding of the impact of the request they are making in terms of additional work – and even, in extreme cases, how what might appear to be a relatively minor change may have a detrimental impact on other aspects of the development.

For example, adding an additional data field to customer details may not seem particularly problematic. The reality, however, could be that doing so would affect data structures far beyond this entity. For example, on every form and screen where the data was displayed, the code that drew the detail from the customer file would need to be modified, and the code that generated any reports might also need to be changed to accommodate the additional values.

Accommodating these changes would be an example of scope creep. This in itself is not an issue, particularly when you remember that methodologies such as Agile and Scrum (covered later in this chapter) are built on this concept as part of normal development activities. However, the reality is that if the level of scope creep is not carefully managed, there will be two potential outcomes:

- The deadline may not be achieved (unhappy client).
- Your profit will be reduced because you are doing more work for the same money (unhappy senior managers).

To avoid scope creep, all team members involved in a development project should be alert to ensure that additional client requests do not derail the project. However, this does not mean that you should immediately dismiss any additional requests made by the client. Ideally, as a development team, you will have discussed at the outset how you would respond to any client requests that are not specifically included in the requirements specification.

One well-used strategy is to raise an invoice for all but the most minor requests. This outlines the additional cost that would be incurred by the client for the inclusion of the requested changes, together with a revised delivery date for the project. It is then up to the client whether they want the amendments to go ahead or not. If not, the organisation simply raises a credit note to cover the invoice.

On a positive note, accommodating scope creep can be one way to produce extra revenue! The outcomes are then:

- The client gets the changes they want (happy client).
- The organisation invoices for the additional work (happy senior managers).

The conflicting demands of stakeholders

Never underestimate the impact that the conflicting demands of stakeholders can have on the outcome of a development project. It is essential that any conflict is eradicated as early as possible in the project. This may not be easy and may require a number of meetings involving heated debate before agreement is reached. The stakeholders should not only agree the requirements and boundaries but also the priority of the components, timelines and so on of a project.

As a software developer, you should not attempt to influence this process. However, you should be available to answer any questions and provide technical clarification about what is or is not possible so as to help the stakeholders to reach a consensus.

Missing deadlines and the ramifications of late delivery

Missing deadlines not only impacts the client but can also affect the team members by increasing the amount of stress they experience when they realise that the project is falling behind. This is particularly the case because missed deadlines tend to create a domino effect where one late project reduces the time available for the next project. This applies additional pressure and may ultimately cause instances of team member sickness.

The outcome could be staff cutting corners, producing a poor development outcome which damages the organisation's reputation. This could affect revenue going forward – not only the remuneration expected at the end of the project but also by reducing the organisation's chance of securing further work if the failure is made public.

An equally significant potential ramification comes in the form of financial penalties, which may eliminate profit or even result in the organisation delivering the project at a loss. There are certainly documented cases of organisations being destroyed by penalties they have been unable to pay, leading to their demise.

Monitoring the development project is essential to ensure that it stays on track. The activity schedule should be revisited regularly to check milestones are being met, adjusting timelines if required and keeping the client informed if relevant. This is why it is essential to build recovery points into the plan to ensure that there is wiggle room if needed.

Tips for meeting software development deadlines

- Review scheduling estimates regularly and before their associated tasks are finalised, taking into consideration input from the development team – unrealistic or overly optimistic estimates are generally not helpful.
- Promote good communication and teamwork – whether via pair programming (see Chapter 7), exchange of ideas or consulting a more senior developer for assistance.
- Get feedback in a timely manner – this will encourage progress and help in meeting future deadlines.

- Be mindful of the larger picture – what are the consequences of *not* meeting the deadline? Damaged reputation? Financial loss? Other consequences?
- Celebrate success – it is important to acknowledge good performance and successful outcomes in software development projects.

Other scheduling issues

In addition to the key issues already mentioned, projects can be impacted by many other less predictable issues, such as:

- **Tasks were significantly more complex** than originally predicted. This also includes the notion of 'gold plating', which is where developers spend too much time adding flourishes or enhancements to components which were neither requested nor expected by the client.
- **Planning was incomplete** or ineffective, leading to a lack of co-ordination.
- **Estimators were inaccurate** in relation to budgets, skills needed or timeframes. The most commonly encountered issue in this area is overly optimistic schedules. In addition, there will be times when you are required to use personnel whom you may feel are less qualified or experienced than you would like due to resource restrictions.
- **Team members may be unexpectedly unavailable** due to illness or bereavement. While annual leave can be considered in the planning and scheduling of activities, it is impossible to predict many other events. Furthermore, key project members may leave to go to other jobs, taking their skills and experience with them.
- **Team co-operation, motivation and/or communication may break down.** This may be caused by the introduction (or removal) of team members, which can negatively influence the team dynamic, or by tension or conflict between team members.
- **Team member performance will vary** over the course of the project and some individuals may let all or part of the team down.
- **Initial design issues may be overlooked or ignored**, and this can have an impact on coding. This is often the result of a failure to produce documentation which adequately records decisions made, leading to mistakes occurring later in the process which then require revisions further down the line.
- **The project team could be let down by third-party tools** which change between the start of the project and the point when they are integrated, or by promised technologies not being available when required.
- **Commitments to other equally or even more important clients** may mean the organisation needs to pull team members off projects. This is the most difficult event to manage, particularly if those team members have become deeply entrenched in the project and do not want to leave it.
- **Poor quality control** in terms of inadequate testing can lead to an underwhelming user or customer experience.

Not monitoring these issues, or simply identifying them too late, can lead to failure.

Regular team meetings will help with the monitoring process, although the frequency of meetings will be dependent on the level and complexity of the project at any particular point in its activities. When meetings are held, you have the option of involving stakeholders, particularly if some ad hoc decisions need to be taken along the way which require their involvement. This is a more attractive option than sending and receiving endless emails.

The importance of the client journey

It is essential to keep the client informed throughout the development process. Why? Because apart from the fact that they are paying the development costs and it's great customer service to keep them informed, managing the client's expectations is crucial.

A good manager of a software development project manages the client's expectations **actively** rather than **reactively**. Anticipating problems helps in preparing appropriate responses and planning how to communicate them.

Keeping the client informed throughout the development process reduces the chance of having to be the bearer of unpleasant ('we **can't** solve the problem') or unexpected ('we **can** solve the problem but it's going to take much longer') news.

Things to consider as part of your project delivery include the following:

- Report project progress and status on a regular basis.
- Involve the client in feature reviews, particularly within an Agile flow, so that the client is fully apprised of the progress and overall quality of the project.
- Document and explain the known risks that could impact delivery – then they won't be a nasty surprise (if the worst should happen).
- Encourage the client to ask questions.
- Be honest in your dealings and explanations (e.g. identified bugs and known issues).

The importance of including all stakeholders in the review

Make certain that representatives from all stakeholder groups have been included in the review phase. This helps to reinforce stakeholders' view that they have been fully included in the process and may well be instrumental in securing future business.

KEY DESIGN METHODOLOGIES AND LIMITATIONS

A methodology is a set of predefined rules, methods or approaches that can be followed to deliver a desired outcome. Think of it as a recipe. All of the ingredients are gathered together and then the steps are followed to produce the final dish.

Some organisations select and apply a methodology in its entirety. Others choose to loosely follow a methodology, selecting only the most relevant rules and processes to suit the project in hand. In addition, you may well find an organisation has a house methodology which draws on a range of other considerations (e.g. the personal preferences of key staff in a project team).

It would be unusual for a new employee in a development context to ask which methodology was being used as this can look somewhat naive. It might therefore be helpful to know some of the key features and uses of some of them so that you can recognise them when you encounter them.

Methodologies are usually categorised as sequential, incremental or iterative, in relation to their phases and how they are conducted. Additionally, the Agile methodology combines elements of iterative and incremental methodologies.

Bear in mind that, in principle, the SDLC (see above) provides a framework which is recognised as the basis for most of the current methodologies, but the amounts of emphasis on the different components may not be the same, and in some methodologies some parts are largely informal or missing altogether.

Sequential methodologies

Sequential methodologies require each phase of the development to have been fully completed (reviewed and essentially signed off) before the next phase begins. The best known (and most commonly used) sequential model is the Waterfall methodology. Just as a waterfall cannot flow backwards, each of the phases in the Waterfall model must be completed before the next phase begins (see Figure 4.2).

Figure 4.2 The Waterfall model

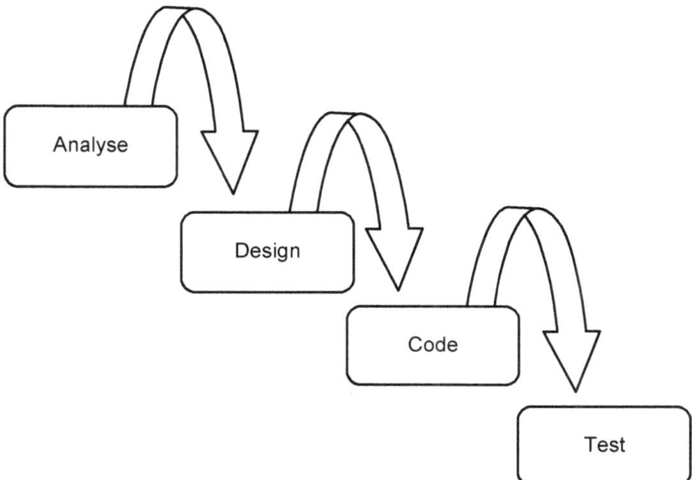

Benefits of the Waterfall methodology include the following:

- Separate, highly defined phases.
- Easy to understand.
- Easy to manage because setting milestones is straightforward.
- Good for predictable developments, particularly where the intended outcome is known in advance.

Drawbacks of the Waterfall methodology include the following:

- Errors in earlier phases are difficult to correct.
- Phases may not overlap.
- The first working version of the software comes late in the development process.
- Clients can feel disconnected from the activity as in the early stages there is little to see.
- Not suitable for complex projects or where innovation would be desirable.

Incremental methodologies

Incremental methodologies support development projects where the requirements of the new system can be broken down into a number of modules where each will become a part of the final working system (see Figure 4.3).

Once each module is complete, it can be released to users.

Figure 4.3 The incremental model

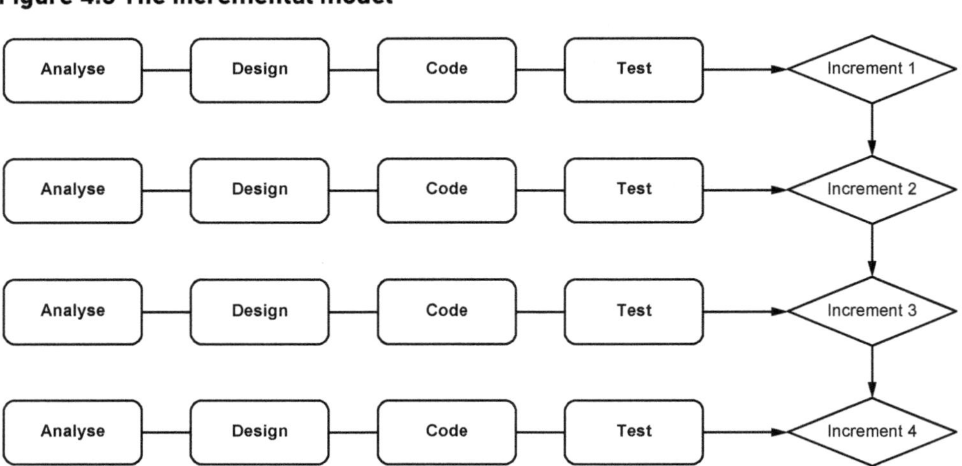

Benefits of the incremental methodology include the following:

- Each phase or increment can be carried out by a separate team, so development can occur in parallel.
- The client will be able to experience the product during development and users will be able to provide feedback.
- It is possible to generate systems quite quickly.
- Any errors will often be found (and resolved) in hours or days, which means that they will not have had the chance to impact too heavily on the development as a whole.
- It can easily accommodate less experienced developers.

Drawbacks of the incremental methodology include the following:

- The ultimate outcomes of the system need to be well understood in advance.
- It needs a good team with a wide range of skills.
- The teams must communicate regularly to ensure that the modules in development will work together when the system is finally pulled together.

Iterative methodologies

Iterative methodologies are based on the quick development of a basic working model: a minimum viable product (MVP) that does the job but that would benefit from refinements either through added functionality or through the incorporation of emerging technologies which provide new possibilities (see Figure 4.4).

Figure 4.4 How to implement the iterative methodology (Source: Kniberg 2016)

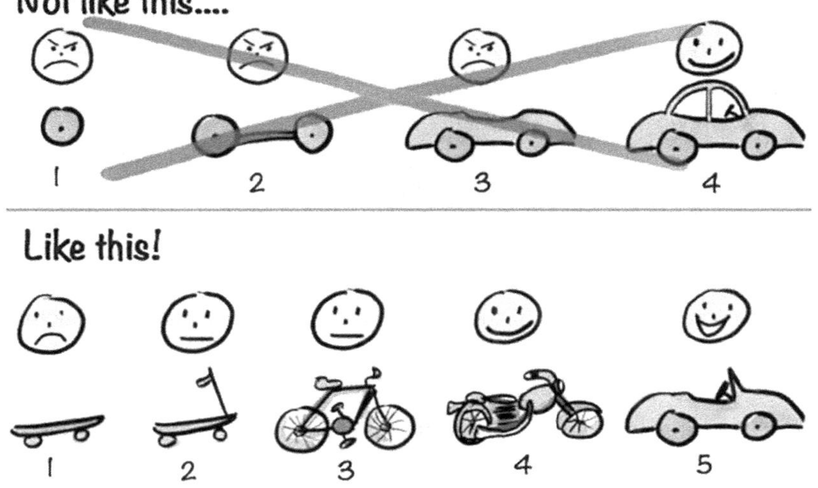

This methodology is ideal where a project may require a significant number of refinements during the development process. This is likely to be because the general direction of travel is known but the specifics of the final solution are less well known.

Benefits of the iterative methodology include the following:

- It allows for innovation.
- It can easily accommodate changes in requirements.
- It is a responsive model, as any client requests can be considered (although it is important to be careful that the project does not become derailed by scope creep; see earlier in this chapter).
- The client is able to feel involved as they can be invited to give feedback.

Drawbacks of the iterative methodology include the following:

- It requires a high level of team interaction at all times.
- The team must be fully committed to ensure success.
- Members of the team must be experienced.
- It can suffer from scope creep.
- It requires a high level of regular interaction with the client and users to check everybody has a shared understanding of what has been agreed.

Agile

Agile, and its derivative Scrum, are classified as both iterative and incremental. These are the approaches that are most commonly used in the IT industry. To be involved in this type of development, it is essential to understand the terminology used. The key elements are as follows:

- **Story:** a story is a top-level overview of a user requirement. It is sufficiently detailed to enable a group of developers to produce an estimate in relation to time and cost.
- **Story points:** these are used to measure the complexity of the story in terms of effort, risk and difficulty (see Chapter 13 for more on these).
- **Epic:** this term describes a larger section of work, usually made up of a number of stories around a particular component or function in the development project.
- **Scrum:** drawn from rugby, this term is used to describe the participants (or team) who will contribute to the development process. The Scrum is made up of a group of individuals who have the required range of skills, such as coding, analysis and testing. Each Scrum has a Scrum master, who essentially leads the activity.
- **Sprint:** an intense (but usually short) period of activity usually lasting about two weeks. The activity is iterative.
- **Sprint planning:** in order to use sprint time effectively, the stories are broken down into specific activities and tasks during sprint planning.

- **Scrum board:** this is used to map and monitor the progress or status of the tasks in the sprint. Some development teams use physical boards to represent the activity, whereas others do this electronically.
- **Sprint demo or review:** this follows a sprint and provides an opportunity for developers to share what they worked on during the sprint, raising the level of awareness among the whole team about the status of the development project.
- **Stand-up:** this is a daily event where the development team gather around the Scrum board (or access an electronic version on a PC or other device) to report back individually on the previous day's work and the plans for the day's activity. The idea is that standing up ensures that this is a relatively short activity lasting only 10 to 15 minutes.
- **Retrospective:** this follows a sprint and provides an opportunity for the Scrum members to evaluate the sprint, its activities and its outcomes. Furthermore, lessons learned from the sprint can carry forward into the next.

The Agile Manifesto

Written in 2001 by 17 software developers, the Agile Manifesto is centred around four key values:

> **Individuals and interactions** over processes and tools.
>
> **Working software** over comprehensive documentation.
>
> **Customer collaboration** over contract negotiation.
>
> **Responding to change** over following a plan.
>
> (Agile Manifesto Authors 2001)

For each statement, the item on the left has more intrinsic value than that on the right. For further information on Agile, refer to *Agile Foundations* (Measey et al. 2015).

HOW IS A DEVELOPMENT METHODOLOGY CHOSEN?

Deciding on the right methodology to use in a development project depends on the type of development being undertaken. Take the example of a new stock system for an online retailer. You would have to decide which model is likely to be the most suitable and beneficial when applied to the development. Would Scrum be a suitable model? Should a sequential model be used instead?

You may find it useful to employ a series of established criteria to help you make the decision. Examples of potential selection criteria are detailed in Table 4.2.

When it comes to choosing the methodology to support a development project, there is not always a clear winner – just some obvious wrong answers. As suggested earlier in this chapter, it can also come down to the personal preferences of the manager (what they know) or other key stakeholders (what they have experienced), or even just the preferred in-house approach (what they do).

Table 4.2 Selection criteria for methodologies

Selection considerations	Description
Required outcome	Some models are more suited to a particular type of development than others. The client brief (see next section) is therefore a key factor in choosing a methodology. For example, the development of a high-risk system (large, complex, potentially liable to fail) should be well analysed, planned and documented. Therefore, the most obvious model in this situation would probably be Waterfall.
Timescale	Development projects which need to be completed quickly will not benefit from a structured model which places emphasis on investigation, design and documentation.
Risk	The development of a data-driven system, such as a stock control system, is likely to be relatively low risk so from this perspective most models would be suitable. In comparison, a development project that involved creating software to automate movement (such as self-driving cars or flight or rail control systems, where errors could result in death or destruction) would benefit from a highly documented, structured methodology to ensure that appropriate analysis and design have been undertaken, all risk factors have been analysed and mitigated, and a number of options have been considered before any coding activity begins.
Accessibility for client and/or users	Methodologies that require a high degree of user interaction may be less suitable for some types of development. A stock system is likely to have future users with a wide range of IT skills. It may therefore be beneficial for a selection of users with varying skills to be involved throughout development, particularly during the design and development stages, to ensure that the system that is produced is accessible to all.
Stability of requirements	It can be argued that a requirement for some tweaks will almost always emerge during a development project, but developments where innovation is preferred do not lend themselves to models that are sequential.
Constraints	A requirement for structure and documentation affects the speed with which a development project can occur. Another example constraint is a lack of experience of a particular methodology (such as Scrum). Constraints are covered in more detail later in this chapter in the section 'System definition'.

HOW THE CLIENT BRIEF AFFECTS THE DEVELOPMENT PROCESS

Each project has unique features, sculpted by the needs of the client and the technical challenges these needs present. It is important to know the required outcome, the timescales involved, the potential risks to the project, the accessibility needs of clients and users, the stability of the requirements and any constraints (see Table 4.2) before you can even begin to understand the problem itself. You can then begin to ask, 'What type of problem are we facing and which skills will be needed to solve it effectively?'

Assume from this point that almost everything that follows applies whether your client is internal or external. This is because an internal client should be treated in much the same way as an external client.

All software development projects are triggered by one or more of the following, which give the development a context:

- identified problem or problems with an existing system;
- changing needs of the organisation;
- business drivers such as reducing costs or improving efficiency;
- the availability of new technologies;
- activities of competitors;
- identification of emerging markets.

In terms of development, the solution or solutions could fall into any one of the following categories or could even need a combination of the solutions detailed below:

- **Off-the-shelf (OTS):** a solution you can simply buy, install and use – for example, an accountancy package or creative software for editing photographs or video.
- **Bespoke:** software that is created completely from scratch to meet a client's needs. This is often necessary when there is no similar product on the market. Bespoke software is often created for a specific context.
- **Tailored:** a solution where an existing product is adapted to include additional functionality so as to make it more suitable for the client.
- **Apps:** essentially, these are software applications which are designed to perform limited or specific tasks. For example, entertainment apps allow users to access services such as Netflix or YouTube, or wider media services such as IMDb, where they can view an online database of TV, film and video game information.
- **Games:** computer and video games are developed for PCs and for devices such as tablets and phones, or they can be played online via a web browser.
- **Hosted services:** those where a third party is paid to provide and manage a range of services.

Let's explore that last one in a little more detail.

Hosted services

There are many types of hosted service, all of which have software products at their core. At the top level, these hosted services fall into one of four categories, as shown in Figure 4.5.

Figure 4.5 Service models

All system components on-site	Infrastructure as a service (IaaS)	Platform as a service (PaaS)	Software as a service (SaaS)
Organisation manages:	***Organisation manages:***	***Organisation manages:***	***Service provider manages:***
Applications	*Applications*	*Applications*	Applications
Data	*Data*	*Data*	Data
Run-time libraries	*Run-time libraries*	***Service provider manages:***	Run-time libraries
Virtualisation	***Service provider manages:***	Run-time libraries	Virtualisation
Servers	Virtualisation	Virtualisation	Servers
Storage	Servers	Servers	Storage
Networking	Storage	Storage	Networking
	Networking	Networking	

Software as a service (SaaS)

SaaS is where an application is distributed and licensed for use over the internet. This means that the software is not downloaded or installed onto the user's machine and both the software and the files created on it can be accessed and used from any computer, providing there is an internet connection.

SaaS provides hosting for applications and development tools and provides a level of database management and business analytics. Examples of SaaS include services such as Dropbox, GoToMeeting, Google Apps, Google Docs and OneDrive. These are useful services for small organisations with limited resources. They also provide opportunities for sharing content during short-term collaborations, particularly where applications may need to be accessed using both web and mobile interfaces.

Platform as a service (PaaS)

Built on virtualisation technology, PaaS provides customers with the services that enable them to develop and run their own software without needing to buy and maintain the usual infrastructure associated with these activities. The customer provides the applications and the data while all of the other services are provided by the third party (including servers, storage, networking, run-time libraries and even the operating system). Examples of PaaS include Google App Engine, Apache Stratos and Windows Azure.

Infrastructure as a service (IaaS)

IaaS allows customers to provide almost all of the software components of the system, with the third party providing the hardware, networking, servers and storage. Examples of IaaS include Google Compute Engine, Amazon Web Services and Rackspace.

Amazon and Google both offer extensive cloud computing services that provide an increasingly wide range of services and technologies to customers, including:

- business applications;
- customer engagement;
- relational databases;
- developer tools;
- IoT (Internet of Things);
- networking and content delivery;
- storage (short-term and long-term, i.e. cheaper 'deep freeze');
- streaming game technology;
- additional cloud-based game processing;
- streaming media services;
- migration and transfer;
- APIs (application programming interfaces).

System definition

Once the client brief has been understood and the reasons for the software development project have been identified, a system specification is generated which defines key aspects of the development. Typical activities include:

- **Identifying key deliverables:** these are the expected goals or outcomes of the development project. The key deliverables will become the criteria for success against which the completed development will be judged.
- **Confirming scope:** this is an essential part of the system definition process as it sets the boundaries for the development project, confirming what will not be included in development as much as what will be.
- **Agreeing timelines:** it is important to always ensure that timelines are realistic. Over-promising and under-delivering can have a whole set of consequences from financial penalties to loss of reputation.
- **Agreeing other milestones:** depending on the development project and the needs of the client, there may be some additional milestones which will need to be observed, such as the delivery of a prototype.

- **Confirming available resources:** the development team must establish the resources they will have at their disposal for the project. This should include consideration of:
  - time;
  - budget;
  - hardware;
  - software;
  - team skills and team availability – in this regard, some mitigating actions may be taken if time allows, such as sending a team member on a relevant course to improve their skills.
- **Understanding constraints:** many people confuse the term 'constraints' with the boundaries or scope of the development project and/or the availability of resources. Instead, in this context, constraints are factors which influence the development and which are essentially non-negotiable, such as complying with particular legislation, having to create a system which will work with an organisational process that cannot be changed, or having to work with existing data capture methods.

The importance of an audit trail

Documenting the key parts of a software development project is essential. This will:

- Provide a record of the decisions taken, together with the rationales.
- Enable the development team to review the phases of the development and to identify what went well or what went badly, and why.
- Allow team members to learn from successes or failures in the process.

Matching client needs with team skills: a team leader's responsibility

It is highly unlikely that an entire development team will be involved in the discussions with the client. In fact, it is possible that some of the team members will never meet the client at all.

The team leader will have had significant contact with the client in the early stages and will have been responsible for producing the system specification. This will have been agreed and signed off by the client as a true and accurate reflection of the requirements of the development. This document is used to focus the activity and is frequently employed as a reference point to ensure that the development project is proceeding in the right direction.

It is important to ensure that the right skills are there when needed. To do this, it is essential for the team leader to know what skills are available to them and what to do if any skills are lacking (e.g. buy in additional expertise or ask team members to undertake training or mentoring, which grow the organisation's future talent).

Very few organisations have the skills they need at their disposal all of the time. If they did, then there would be no need for training or continuing professional development programmes (in which staff are encouraged or required to improve their existing skills or develop new ones).

Understanding development requirements and best practice

Once the client brief has been crafted as a requirements specification, the nuts and bolts required to support the development should be put in place. When it comes to software development, there are several key requirements:

- Availability of appropriate resources, such as hardware, software, diagramming tools and development platforms, as required by the project.
- Selection of team members, making sure that there is the right mixture of skills and experience. The team should usually include some less experienced team members so that they can be given an opportunity to develop.
- An extensive plan that allows for all activities required for the project, allocated to team members as needed.
- Identification of milestones to ensure that the project can be suitably monitored.
- Identification of recovery points, which build in flexibility and make it more likely that the deadlines will be met.
- Agreements about the version control system that will be used to manage the documentation (including the code).
- Agreement on the naming of functions, variables and so on.
- Standard coding conventions (e.g. bracing styles, if required).

SUMMARY

In this chapter we have examined various software development methodologies and situations that can arise during the SDLC (both good and bad). We have considered the client brief and how this is fashioned into a requirements specification which will form the basis for the development.

In the next chapter we will examine the various commercial software languages available.

5 OVERVIEW OF COMMERCIAL SOFTWARE LANGUAGES AND PARADIGMS

'There are only two kinds of languages: the ones people complain about and the ones nobody uses.'

– Bjarne Stroustrup

Being an effective software developer doesn't necessarily mean having **all** the answers but having enough knowledge and understanding to contribute sensibly and effectively. Having an understanding and appreciation of commercial languages and development paradigms will enable you to make those contributions.

TRENDS IN PROGRAMMING LANGUAGES

There are many programming languages in existence; the TIOBE (The Importance of Being Earnest) index (https://www.tiobe.com/tiobe-index) tracks and measures the popularity of hundreds. The Language List (originally posted to the Usenet group comp.land.misc in the early 1990s) collected information on over 2,000 computer languages and Wikipedia lists just over 700 at the time of writing. Realistically, the number may exceed tens of thousands, and the exact number will probably never be known.

Not all of them are in active development or even still commercially used. The growth, popularity and influential spread of certain programming languages have shaped the modern software development industry. A good example is Haskell – although it is not at the time of writing in the top 20 programming languages according to Google searches (based on the Popularity of Programming Language (PYPL) index; see http://pypl.github.io/PYPL.html), it has influenced the development of more popular languages, such as Python and C++.

Programming languages can be categorised in several different ways. A popular approach is to consider their paradigms – their style or way of doing something. Common paradigms include:

- **Procedural:** solves problems in an algorithmic way, with the code typically broken down into procedures (e.g. functions).
- **Object-oriented:** models the real world, creating encapsulated classes which interact with others to create solutions.
- **Declarative:** defines what is needed but not how this is to be achieved.
- **Functional:** solves problems using stateless functional components.
- **Event-driven:** captures system or user events and writes suitable handlers to deal with them.

This list is **not** exhaustive: many different variations and combinations exist. Broadly speaking, programming languages tend to get (rightly or wrongly) pigeonholed into a single paradigm, for example:

- C is procedural.
- C++ and Java are object-oriented.
- Prolog is declarative.
- Visual Basic .NET is event-driven.

However, the distinction is not always clear-cut. For instance, Java also supports a user's interaction with an object, such as clicking a button; this is a classic trait of an event-driven feature. Python, a very popular language at the time of writing, supports procedural, object-oriented and functional programming paradigms.

Fortunately, many development tools (editors, compilers, debuggers, etc.) are freely downloadable, offering the fledgling programmer many opportunities to learn new skills and experience different approaches to problem-solving prior to gaining employment and working on commercial projects. The key question is: which programming languages should you focus on?

The simple (if obvious) answer is: the ones that are in commercial demand. After all, you want to start a successful and financially rewarding career as a programmer, don't you?

So, which ones are popular? Well, that depends on who you ask, so let's explore a few options.

The recruitment perspective

When examining programming languages from a recruitment perspective, we are essentially looking at the skills for which the IT industry is advertising. This is helpful as it lets us know which skills are in most demand and which are growing and shrinking in popularity.

As in all things, supply and demand tend to determine salaries and working conditions. The UK IT industry often finds itself in short supply of certain talents as technologies quickly evolve and professional development fails to keep track. Keeping ahead of the development curve is therefore a good idea. Figure 5.1 gives an overview of what's actually in demand.

If this data is representative, we can make some quick observations:

- Java and Python appear to be the most popular language choices.
- Demand for most of these 'top' languages is increasing.
- Demand for Perl is noticeably decreasing (anecdotally, being replaced by Python).
- Demand for Python is bucking the overall trend by increasing year on year.

Figure 5.1 Programming language trends based on numbers of jobs posted worldwide on Indeed.com (Source: Dowling 2020)

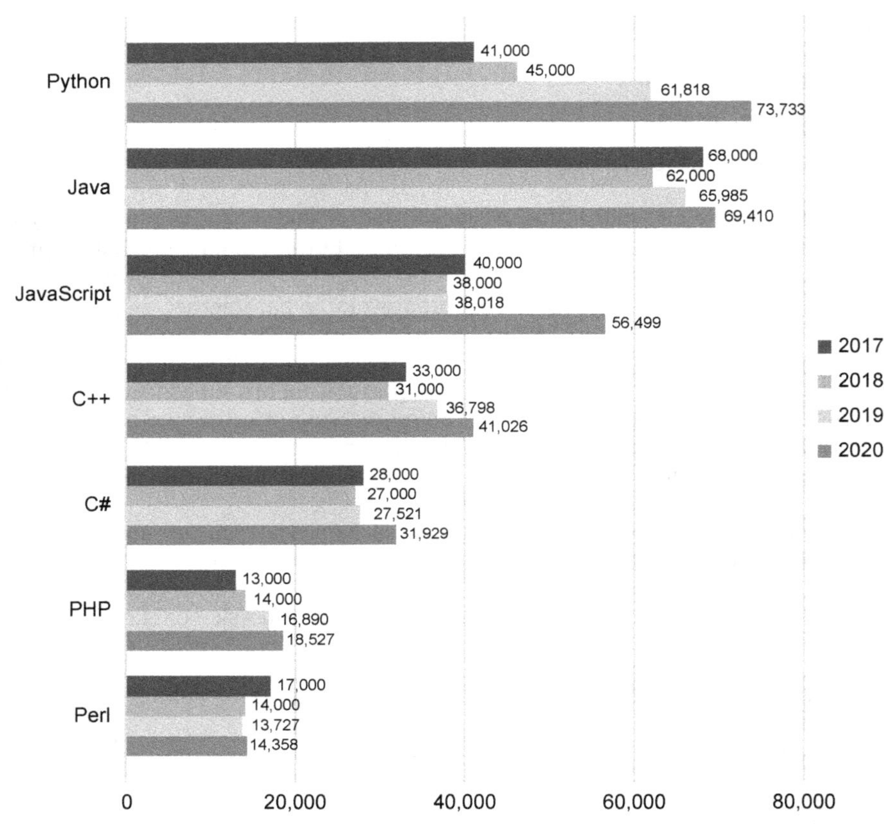

The growth in the popularity of languages such as Python is not surprising. It has been warmly embraced by the scientific community, has proven adept as a data-mining tool, and is popular in data science and machine learning applications given its rich and diverse number-crunching libraries.

In contrast, many jobs previously tackled using Perl, such as automating operating system processes or gluing together parts of an enterprise computing solution, are inexorably shifting to Python. Perl and Python have many similar features and are both available on multiple platforms (including Apple macOS X, Microsoft Windows and Linux). Arguably, though, the Python programming language is more readable and, consequently, much more maintainable. This is always a consideration, as is the growing reliance on Python for data science and machine learning applications.

PHP and JavaScript form the two-chamber heart of most commercial websites (offering server- and client-side scripting respectively), so their inclusion and enduring popularity are not surprising. Additionally, JavaScript is gaining greater importance with the growing popularity of the Angular and React frameworks. C#'s popularity may also be partly attributed to its use in building ASP.NET applications.

Languages such as C# and C++ (including C) are extremely portable languages – general purpose and therefore ubiquitous in many aspects of IT service development. Their popularity will fluctuate but, as with Java, they are far too entrenched not to be considered strong favourites in career development and ongoing project development.

Periodically, claims will be made in the IT sector that a popular programming language is 'dead'. Much like the unsubstantiated claims about the author Mark Twain in the 1890s, their demise is also over-reported, exaggerated or simply being used as cheap click-bait on technology-based news sites.

A good example of this is PHP, a mature though often disparaged language that (as mentioned) is commonly seen as the workhorse of the server-side scripting part of a full-stack web solution, particularly in shared hosting and cloud-based environments. For many years, WAMP (Windows, Apache, MySQL and PHP) or more typically LAMP (Linux, Apache, MySQL and PHP) software bundles formed the technology stack of many commercial websites. The reasons for this were clear:

- All the components are well known, well documented (incredibly important) and (generally) well respected:
  - Windows/Linux: operating system;
  - Apache: HTTP web server;
  - MySQL: multi-threaded, multi-user relational database management system;
  - PHP: server-side scripting language.
- All of the components are free.
- They integrate very well.
- They are generally reliable, both separately and when used together.

However, from a commercial perspective, although many other options exist – for example, using NGINX instead of Apache and PostgreSQL instead of MySQL, and using Node.js as a server-side scripting component – recruitment opportunities for PHP developers remain very buoyant at the time of writing. Granted, they may require additional skills (e.g. knowledge of a modern PHP framework, such as Laravel), but new projects are still being created using this now 'aged' language. Why? The sector simply has a lot of **experienced** PHP programmers, many of whom are in the middle or senior bracket. In other words, the potential recruitment options for a commercial PHP solution are numerous, broad and **skilled**.

How many sites use PHP? Is it a language to consider as we move into the third decade of the 21st century? It is well documented that the evergreen content management system (CMS) WordPress (its core and plug-ins) is mostly based on PHP and MySQL. At the time of writing, WordPress has a CMS market share of around 61%, representing around 37% of all websites (Osman 2020). That's a large potential employment market to target.

The version control system perspective

GitHub, a company at the very heart of the open source software development drive, has a special interest in tracking programming trends. Because it exists as a hub for millions of developers and organisations, its annual top-ten overview of languages often makes interesting reading (see Figure 5.2 and Octoverse 2019).

Figure 5.2 Top languages according to GitHub, 2014–2019 (Octoverse 2019)

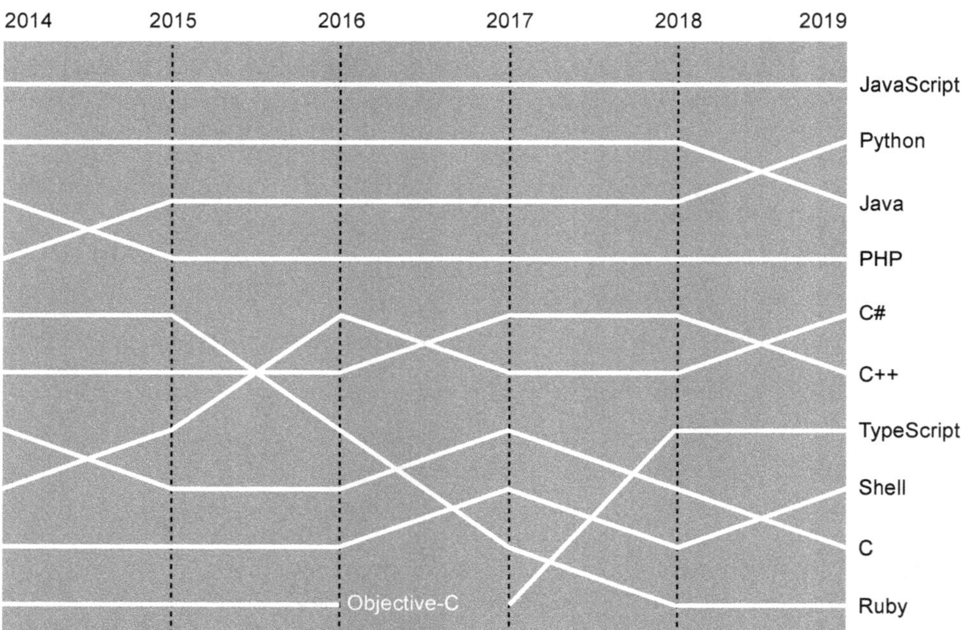

From a simple analysis of changes in the popularity of these languages over time, we can make the following observations:

- There has been very little change over the past six years for the four most popular languages:
  - JavaScript;
  - Python;
  - Java;
  - PHP.
- C projects are migrating (possibly to C++ and C#).

- Objective-C, once a mainstay of Apple macOS and iOS development, has seemingly reduced in popularity due to Apple's development of the Swift programming language in 2014.
- Shell scripting remains an important part of a programmer's toolbox.
- Microsoft's TypeScript is one of the fastest-growing languages. It is a superset of JavaScript and offers optional static typing, classes and interfaces, all of which address many of the perceived weaknesses in the JavaScript language.

The Popularity of Programming Language (PYPL) index perspective

As introduced earlier in the chapter, the PYPL uses raw data from Google Trends to track the popularity of programming languages across the world based on the frequency of search terms. The current worldwide ranked view (as of May 2020 compared to a year previously) is shown in Table 5.1. We can see that many of the languages previously discussed are present and accounted for in this list. Although the list doesn't map to commercial usage, it does reflect languages of interest, and that is a broad indicator of popularity in terms of learning, academic research and implementation.

Table 5.1 Popularity of programming languages in May 2020 compared to a year ago (Source: PYPL Index, http://pypl.github.io/PYPL.html; this work is licensed under a Creative Commons Attribution 3.0 Unported License (CC BY 3.0))

Rank	Change	Language	Share	Trend
1		Python	31.17%	+4.3%
2		Java	17.75%	–2.4%
3		Javascript	7.99%	–0.3%
4		C#	7.05%	–0.2%
5		PHP	6.09%	–1.0%
6		C/C++	5.67%	–0.3%
7		R	3.93%	–0.1%
8		Objective-C	2.40%	–0.4%
9		Swift	2.26%	–0.1%
10	↑	TypeScript	1.89%	+0.3%
11	↓	Matlab	1.81%	–0.2%
12	↑↑	Kotlin	1.55%	+0.3%
13		VBA	1.33%	+0.0%
14	↑	Go	1.26%	+0.1%
15	↓↓↓	Ruby	1.19%	–0.2%

The programming community perspective

The active programming community often offers clues about the popularity of languages. Programming queries logged through popular question-and-answer websites (such as the evergreen Stack Overflow) tend to suggest the respective levels of use of different languages in the sector. Here is a ranked list of the most popular languages according to a survey carried out by Stack Overflow (2019):

1. JavaScript
2. HTML/CSS
3. SQL
4. Python
5. Java
6. Bash/Shell/PowerShell
7. C#
8. PHP
9. TypeScript
10. C++

Although this list (filtered to target just professional developers) also includes various web-development-related technologies – such as HTML, CSS and SQL – it should be noted that, broadly speaking, the same choices can be seen. Once again, the popular development languages include C++, C#, Java, JavaScript, PHP and Python.

Trends in programming languages: conclusions

At the time of writing, many languages are growing in popularity, such as Python and TypeScript, while other languages remain entrenched as go-to solutions for professional developers, such as C++, C#, Java and JavaScript. Conversely, once-popular languages such as Perl are now falling into decline, despite ongoing development on the language by its dedicated teams.

The picture may seem confusing at first, but it is always worth considering the nature of the languages involved. Historically speaking, many of these popular programming languages have inherited the syntax, if not necessarily the 'spirit', of older languages such as C. Popular choices such as C++, C#, Java, JavaScript, PHP and Python all contain many building blocks which a trained C programmer would instantly recognise. As such, although many different languages exist, they have many similarities as well as key differences, and such similarities can help to popularise a new language in a crowded space.

New languages will continue to be developed, especially as real-world computing needs continue to diversify and change. Continuing professional development is therefore essential for a programmer, as there will always be new approaches and new ideas to learn, absorb and apply. Keeping your technical skills up to date is vital, so always take the opportunity to do so when you can.

LANGUAGE BUILDING BLOCKS

Every programming language has several elements that you will use to build solutions. It is beyond the scope of this book (and almost an impossible task) to cover the syntactic detail of all languages, so a checklist of core features follows, accompanied by sample syntax from popular languages where appropriate.

Data types

Most languages have defined data types, such as **`char`**, **`integer`**, **`float`** and **`string`**. You must ensure that data can be safely stored in your chosen types, especially for numeric types (i.e. the data must not be too small, or so large that it overflows).

Atomic data items are the lowest level of detail that a computerised solution might use, because they are not usually broken down into smaller chunks. For example, although an application might store a customer's address and telephone number, the address could be broken down into a number of address lines, including house number, street and city. In contrast, the customer's telephone number is essentially a single item of data. The most common types of atomic data we can use are shown in Table 5.2.

Table 5.2 Data types with examples

Data type	Description	Example values
`Integer`	A whole number with no fractional part	**`0, 100, -400`**
`Real`	A number which can contain a fractional part	**`0.0, 1.5, -400.99`**
`Char`	A single character, usually delimited with single quotes	**`'a', 'B', '@'`**
`String`	0 or more characters, usually delimited with double quotes	**`"Hello World!", "A", ""`**
`Boolean`	A logical value, named after the English mathematician George Boole	**`TRUE, FALSE`**
`Date`	A valid calendar date	Depends on the format used in the country. For example, **`10/07/2020`** is 10 July 2020 in dd/mm/yyyy format or 7 October 2020 in mm/dd/yyyy format

Selecting appropriate data types is crucial to a project's success. A customer's telephone number might be 11 digits long and could be stored as an **`integer`** but, as a rule, it is more likely to use a **`string`** as no arithmetic will need to be performed on it.

Input and output (I/O)

Programming languages typically need methods to both accept input **from `stdin`** (the keyboard) and send messages **to `stdout`** (the screen). Additional interface options can be created using visual components to build mouse-controlled GUIs or leverage HTML and CSS to build client-side interfaces within a web browser.

No matter the language, the syntax is usually quite similar, as shown in Table 5.3.

Table 5.3 Comparison of input and output statements

	Language	Statement
Output	C	`printf("Hello World!");`
	Python 3.X	`print("Hello World!")`
	Java	`System.out.println("Hello World!");`
Input	C	`printf("Enter username: ")` `fgets(name, 30, stdin);`
	Python 3.X	`username = input("Enter username: ")`
	Java	`Scanner myObj = new Scanner(System.in);` `System.out.println("Enter username :");` `String userName = myObj.nextLine();`

Care should be taken to consider I/O functions which are no longer secure and subject to buffer overflow attacks, such as the **`gets`** (get string) function in C. Many others exist!

Operators

Most languages make use of special symbols that perform arithmetic, logical operations or relational operations on data – these are operators. Although there is a high degree of commonality, these can vary from language to language, so care is required. Sample operators may include those shown in Table 5.4.

Operators can also be defined as being unary, binary or (in some languages) tertiary, depending on the number of operands they have, for example:

- **`-12`** (a unary subtraction operator, indicating a negative number, only acting on one operand);
- **`12 - 3`** (a binary subtraction operator, subtracting 3 from 12 as it acts on two operands).

It should also be noted that operators in any programming language follow an order of precedence, a superset of the BODMAS (brackets, orders, division, multiplication, addition, subtraction) or BIDMAS (brackets, indices, division, multiplication, addition, subtraction) rules you may have learned in maths classes at school. For example, in a simple mean average calculation:

$$average = (num1 + num2 + num3) / 3$$

Table 5.4 List of generic operators

Category	Operator	Description
Arithmetic	+	Addition
	-	Subtraction
	*	Multiplication
	/	Division
	MOD %	Modulus (remainder from integer division)
Logical	&& AND	Boolean **AND** operation
	|| OR	Boolean **OR** operation
	! NOT	Boolean **NOT** operation (inverter)
Relational	>	Greater than
	<	Less than
	>=	Greater than or equal to
	<=	Less than or equal to
	= == ===	Test for equality (equal to)
	< > != !==	Test for inequality (not equal to)

As you can see, parentheses (brackets) are used to change the order of arithmetic operations in this code sample. Without it, *num3* would be divided by 3 **first** as the division operator has **higher precedence** than the addition (and we want to do all additions **before** dividing). The actual order of precedence depends on the specific language's list of operators, so always familiarise yourself with it in your target language to reduce the chance of errors. When multiple operators exist at the same level of precedence, associativity rules apply and these are usually interpreted right to left or left to right.

Let's examine a simple formula and explore its interpretation and conversion to program code. This formula, written in mathematical notation, is used to convert a temperature from degrees Celsius to degrees Fahrenheit:

$$°F = \left(°C \times \frac{9}{5}\right) + 32$$

Performing arithmetic calculations in pursuit of a working software solution is a common task for a developer. They should be able to read this type of formula and convert it to the target programming language; for example, in Python:

```
deg_f = (deg_c * 9/5) + 32
```

As you can see, the formula has been converted quite neatly into the programming language and, provided sensible data values for **deg_c**, should calculate the appropriate Fahrenheit equivalent.

Of course, the formula may be written somewhat differently:

$$°F = \frac{9}{5}°C + 32$$

Despite this, the resulting line of code would be dutifully converted to the target language. Note that the **lack** of operator between the two adjacent terms: $\frac{9}{5}$ and °C conventionally **implies** a multiplication.

Consequently, the resulting conversion to Python would be:

```
deg_f = 9/5 * deg_c + 32
```

If we test these two Python alternatives in the development shell, we get:

```
>>> deg_c = 100.0
>>> deg_f = (deg_c * 9/5) + 32
>>> deg_f
212.0
>>> deg_f = 9/5 * deg_c + 32
>>> deg_f
212.0
```

As you can see, despite the **different constructions** of the arithmetic expressions, **both statements** return the same result. The following programming and mathematical principles are at play here:

- **Commutative law:** the order of terms in the multiplication doesn't actually matter: **`a*b`** = **`b*a`**.
- **Order of precedence (BODMAS/BIDMAS):** multiplication and division have higher precedence than addition.
- **Operator associativity:** although multiplication and division have the same precedence, their associativity reads left to right (division then multiplication).

Being able to read, interpret and replicate mathematical formulae into a target programming language should be considered a core skill.

Constructs

A programming construct is a building block which is used in the construction of a program. There are three basic constructs (or building blocks):

- **Sequence:** a linear set of statements which execute one after the other in strict sequence, none repeated, none missed (see Figure 5.3).
- **Selection:** a choice, typically an **`if`**…**`else`** or switch-style statement offering the programmer branching (different logical pathways) depending on the result of a Boolean condition (see Figure 5.4).

- **Iteration:** a section of code which repeats while a condition is true. Some are pre- or post-conditioned or are typically set to run a fixed number of times – for example, `for loop`, `while loop` and `do...while` (see Figure 5.5).

Figure 5.3 Sequence

Figure 5.4 Selection

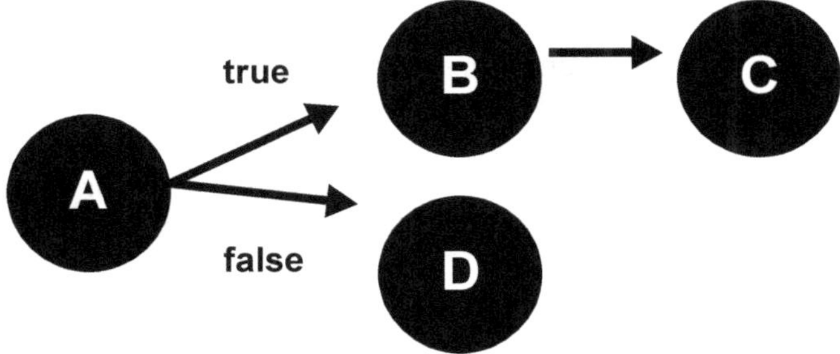

Figure 5.5 A post-conditioned iteration

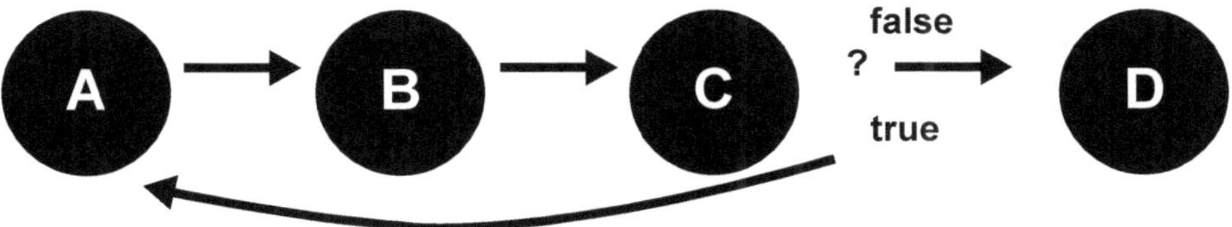

Data structures

Data structures are simply containers that are used to organise loose data items. Many different structures are available, as shown in Table 5.5.

It should also be noted that most data structures can be nested. For example, it is certainly possible to create and maintain a stack containing arrays or a linked list of queues.

Many modern languages have variants of these as built-in data types which require very little additional programming to get up and running. Knowing which data structure

Table 5.5 Types of data structure

Data structure	Description
Array	One-, two- or multi-dimensional lists of values. Traditionally static in size (cannot shrink or grow) and usually containing data of the same type – although this can be a compound data type, such as a record structure.
Record-like structure	A collection of different (but related) data items.
Queue	A FIFO (first in, first out) dynamic data structure where data can be added to the tail and removed from the head.
Stack	A LIFO (last in, first out) dynamic data structure where data can be popped from the stack top after first being pushed onto it.
Tree	A hierarchical data structure which offers more efficient search and storage options than a linear data structure, such as an array.
Linked list	A dynamic data structure, created on the heap, which has no fixed size. Programmers must maintain dynamic links between each of the linked list nodes.

to use (and when) is key to most commercial problem-solving and it is often a wise move to take time in the design phase to identify them correctly; rash choices will come back to haunt you in further iterations of the software when their operational limits and unwanted behaviours and attributes become development blockers.

ALGORITHMS

An algorithm is usually defined as a sequence of steps or operations that are designed to solve a problem.

Common algorithms

Many different algorithms can be used in a program. Common ones that reward memorisation include:

- **input validation:** ensures an inputted value is within a set range;
- **string manipulation:** often used for concatenation (joining strings together) or extracting substrings;
- **linear search:** searches a data structure (typically an array or linked list) for a given value;
- **binary search:** searches a binary tree for a given value;

- **sorting:** organises data into a given order (languages often have library sort functions but the mechanics of popular sorts, including an optimised bubble sort, quick sort and insertion sort, are beneficial);
- **reading from, and writing data to, a file:** often used for reading CSV (Comma-Separated Values files) or appending to log files;
- **recursion:** a function that calls itself to solve a problem – common examples include the Fibonacci sequence, factorials, Newton's square root and any tree traversal (such as a file system directory).

This list is not intended to be exhaustive (it only scratches the surface) but it does provide a foundational layer for a would-be software developer in a competitive field.

Measuring algorithm complexity: Big 'O' notation

Algorithms don't tend to be measured in terms of lines of code. Indeed, in many cases there isn't really an obvious relationship between complexity and line length; recursion is a classic example because although the code is typically short, its tracing and debugging can prove problematic for even the most experienced developer.

A somewhat more useful measure when considering algorithms is the concept of performance – for example, the algorithm's efficiency in terms of execution time and the amount of RAM (random access memory) used. This performance measurement is commonly called Big 'O' notation.

Five classifications exist:

- O(1)
- O(n)
- O(n2)
- O(2n)
- O(log(n))

However, it is the first three that we're chiefly interested in from a programming perspective. Let's examine these with some simple examples.

O(1) algorithm: constant time

O(1) is the most basic classification, describing algorithms which **always take the same execution time** no matter the size of the data set. An example is given in Table 5.6.

In the example in Table 5.7, it should take the same amount of time ('constant time') to access the **first element** of the array, despite the arrays being of different lengths and data types.

Table 5.6 Example array containing `EmployeeNames`

Employee Names	[0]	[1]	[2]	[3]	[4]	[5]	[6]
	K.Ishiguro	B.Dylan	S.Alexievich	M.Yan	D.Lessing	D.Fo	W.Falkner

Table 5.7 Example array containing vowels

Vowels	[0]	[1]	[2]	[3]	[4]
	A	E	I	O	U

O(n) algorithm: linear search

O(n) is a common operation which works by sequentially comparing data items in a list against a given search value until either a match occurs or the end of the list is reached with no match being found. In complexity terms this is designated O(n) – the time taken to search increases with the length of the list ('n').

As an example, you may wish to search an array of employees to find and output their staff ID (the matching array index) or display 'not found'. See Table 5.8.

Table 5.8 Example array

Employee Names	[0]	[1]	[2]	[3]	[4]	[5]	[6]
	K.Ishiguro	B.Dylan	S.Alexievich	M.Yan	D.Lessing	D.Fo	W.Falkner

To solve this problem you could use a combination of a **`FOR`** loop and **`IF`** statement to walk the array from the first to last element and compare each value against an inputted name. If the name matches, the loop's counter value is output (the staff ID) and a Boolean flag is set to **`TRUE`**. If the Boolean flag has never been set to **`TRUE`** after the loop ends, it is safe to say that the search value has not been found.

Suggested pseudocode is as follows (pseudocode is a natural-language-based algorithm which avoids using any particular programming language's syntax; see Chapter 11 for more information):

```
DECLARE EmployeeNames : ARRAY[1:30] OF STRING
DECLARE SearchName : STRING
DECLARE Counter : INTEGER

EmployeeNames [0] ← "K.Ishiguro"
EmployeeNames [1] ← "B.Dylan"
EmployeeNames [2] ← "S.Alexievich"
EmployeeNames [3] ← "M.Yan"
EmployeeNames [4] ← "D.Lessing"
```

```
EmployeeNames [5] ← "D.Fo"
EmployeeNames [6] ← "W.Falkner"

FOUND ← FALSE
OUTPUT "Enter name to search: "
INPUT SearchName
FOR Counter ← 0 TO 6
 IF SearchName = EmployeeNames[Counter]
  THEN
   OUTPUT "StaffID is ", Counter
   FOUND ← TRUE
 ENDIF
ENDFOR
IF FOUND = FALSE
 THEN
  OUTPUT "Sorry, ", SearchName, " was not found."
ENDIF
```

In reality, many programming languages have this type of common functionality built in. For example, in Python, the **`index()`** method can be used to achieve a similar goal without the need for a complex algorithm:

```
>>> employee_names = ["K.Ishiguro", "B.Dylan", "S.Alexievich", "M.Yan",
"D.Lessing", "D.Fo", "W.Falkner"]
>>> employee_names.index("B.Dylan")
1
```

O(n^2) algorithm: sorting

Sorting is a common task for software developers. It involves arranging data items into a sequence based on a criterion – for instance, alphabetically, chronologically (by date) or by value. Items can be sorted in ascending order (smallest to largest) or descending (largest to smallest). Although most programming languages have efficient functions to fulfil most sorting needs, an understanding of the underlying algorithms involved is often instructive.

A bubble sort is a simple (but somewhat inefficient) O(n^2) 'quadratic algorithm' that repeatedly steps through a list in order to reorganise its data into a preferred order. Typically, multiple passes through the list are needed to switch out-of-order pairs until the list is fully sorted. Let's consider a sample scenario and walk through the logic involved, step by step.

To improve a customer's user experience (UX), the search results on an e-commerce website need to display all matched products in ascending alphabetical order. We'll start by creating some sample test data and trace through the passes required to sort these results successfully. Figure 5.6 shows the first pass.

Figure 5.6 First pass of a sorting algorithm

First pass (starts **unsorted**)

StockItems	[1]	[2]	[3]	[4]	[5]
	Tape Measure	Screwdriver	Hole Punch	Steel Ruler	Craft Knife

First pass (**switch pairs**)

StockItems	[1]	[2]	[3]	[4]	[5]
	Tape Measure	Screwdriver	Hole Punch	Steel Ruler	Craft Knife

First pass (**switch pairs**)

StockItems	[1]	[2]	[3]	[4]	[5]
	Screwdriver	Tape Measure	Hole Punch	Steel Ruler	Craft Knife

First pass (**switch pairs**)

StockItems	[1]	[2]	[3]	[4]	[5]
	Screwdriver	Hole Punch	Tape Measure	Steel Ruler	Craft Knife

First pass (**switch pairs**)

StockItems	[1]	[2]	[3]	[4]	[5]
	Screwdriver	Hole Punch	Steel Ruler	Tape Measure	Craft Knife

First pass (**complete**)

StockItems	[1]	[2]	[3]	[4]	[5]
	Screwdriver	Hole Punch	Steel Ruler	Craft Knife	Tape Measure

After the first pass, the stock names still aren't fully sorted. However, the alphabetical last name 'Tape Measure' has floated (like a bubble, hence the name) to the end of the array.

Another pass is clearly needed, though, as this data is not yet fully sorted (see Figure 5.7).

Figure 5.7 Second pass of a sorting algorithm

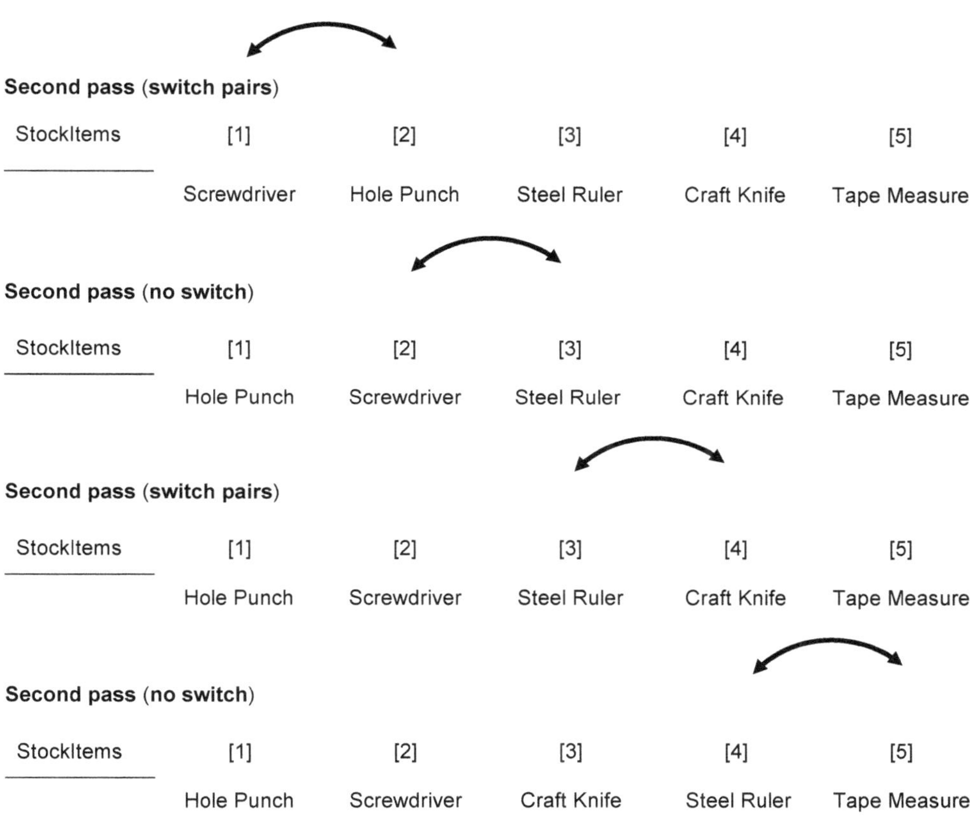

After the second pass, the order is better but still not correct, so a third pass is needed (see Figure 5.8).

You will no doubt have noticed after the third pass that the 'Craft Knife' element is still out of order; we'll need to do another pass (see Figure 5.9).

At this point we can see that the array is fully sorted. However, the standard inefficient bubble sort algorithm **will not stop** until it has made five full passes (as there are five elements). This means that in this case most of the fourth and the entire fifth pass are completely unnecessary.

The bubble sort algorithm is one of the easiest to understand. However it is far from being the most efficient.

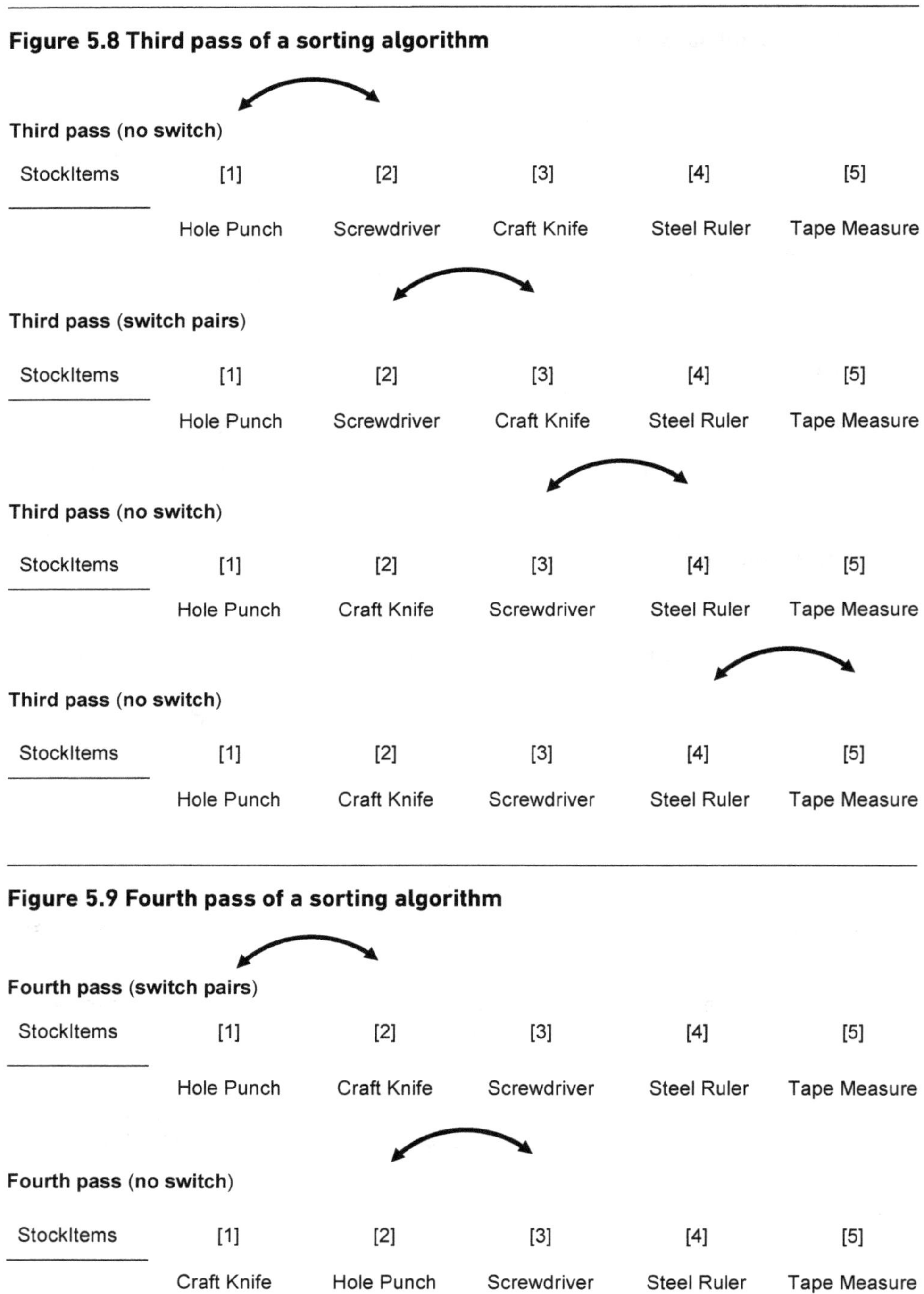

Figure 5.8 Third pass of a sorting algorithm

Figure 5.9 Fourth pass of a sorting algorithm

Suggested pseudocode:

```
DECLARE StockNames : ARRAY[1:5] OF STRING
DECLARE Pass : INTEGER
DECLARE Counter : INTEGER
DECLARE TempString : STRING

StockNames[1] ← "Tape Measure"
StockNames[2] ← "Screwdriver"
StockNames[3] ← "Hole Punch"
StockNames[4] ← "Steel Ruler"
StockNames[5] ← "Craft Knife"

FOR Pass ← 1 TO 5
 FOR Counter ← 1 TO 4
  IF StockNames[Counter+1] < StockNames[Counter]
   THEN
    TempString ← StockNames[Counter]
    StockNames[Counter] ← StockNames[Counter+1]
    StockNames[Counter+1] ← TempString
  ENDIF
 ENDFOR
ENDFOR
```

Again, due to its necessity, sorting functionality is commonly built into most modern programming languages. For example, in Python, one technique is to use the **`sorted()`** function:

```
>>> stock_items = ["Tape Measure", "Screwdriver", "Hole Punch", "Steel
Ruler", "Craft Knife"]
>>> sorted(stock_items)
['Craft Knife', 'Hole Punch', 'Screwdriver', 'Steel Ruler', 'Tape
Measure']
```

It is worth noting that due to the standard unoptimised $O(n^2)$ bubble sort being particularly inefficient (when compared to other sort methods), most built-in sort functions tend to use a more efficient algorithm. Python uses one called 'Timsort', as created by Tim Peters in 2002.

COMMON PROGRAMMING PARADIGMS

Modular programming

Modular programming is a paradigm that encourages the decomposition of a large, monolithic solution into smaller, reusable modules of code which can be individually written and tested by a team of programmers working in parallel, thereby reducing development times. These are variously called 'modules', 'procedures', 'functions' or 'subroutines', depending on the language in question. They commonly consist of between 5 and 50 lines of code. This approach also forms the key concepts of functional programming.

The key advantages of writing code in a modular fashion are:

- improved readability and understanding;
- should follow the single responsibility principle (SRP), where each piece of code only does one thing;
- can be reused in a solution, helping to achieve the DRY (don't repeat yourself) principle;
- can reduce the scope of potential debugging;
- helps a programmer to decompose a large problem into smaller pieces which are easier to solve;
- ideal for team working on a shared project (if design documentation is followed).

Popular modules can be placed in a library and reused in future projects, further amortising the original development time and costs.

Object-oriented programming

Object-oriented programming (OOP) is at the heart of many of the cornerstone languages used in the software development industry. Java, Python, PHP and C++ are all popular languages which have the core object-oriented principles.

For a software developer, understanding the principles of OOP has become a must. In the past it was always an option – now it's almost obligatory to consider it when solving any problem.

It would be difficult in a generalist book to cover this programming paradigm in sufficient depth – but, essentially, this approach models the 'things' which interact in the real world. These things may be physical, such as people, or logical, such as a bank account.

Either way, each of these 'things' has a set of attributes:

- **properties:** things that describe them (their data);
- **methods:** things that they do (their functions or actions).

The act of combining these two attribute types (properties and methods) into a single unit is called 'encapsulation'. When you encapsulate attributes like this, you have essentially created a 'class', as shown in Figure 5.10.

A class acts a bit like a template, allowing concrete 'instances' to be created in its image. This process is called 'instantiation' and these created instances have a more popular name: 'objects'.

Let's put this into context by looking at a very simple example.

Figure 5.10 A class showing encapsulated properties and methods

Imagine a bank account. It exists as a 'thing' and has properties and methods. We can model this using a simple visual tool – a 'class schema' (see Figure 5.11).

Figure 5.11 Class schema

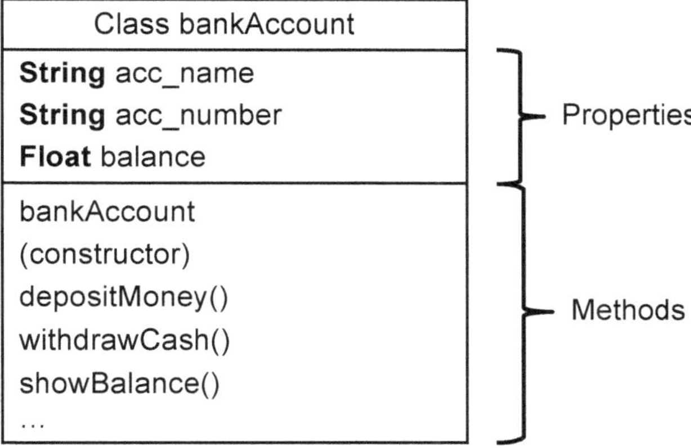

In a class, methods tend to be **public** while the properties remain **private**; this reinforces the concepts of data hiding and controlled scope – that is, it is not possible to access an object's private data other than from the public methods.

Classes have a special method called a 'constructor', which executes automatically when an object of its class type is declared. The constructor is typically used to initialise the private data in the class. 'Destructors' may also be used, being called when an object is going out of scope and its memory is being collected as garbage. Another common technique is 'setters and getters', used to retrospectively set and get private data stored in the object.

We can start the implementation of this class quite simply in a language which supports object orientation, in this case Python:

```
class bankAccount(object):
    """A simple bank account class"""

    def __init__ (self, acc_name, acc_number, balance = 0):
        self.__acc_name = acc_name
        self.__acc_number = acc_number
        self.__balance = balance

    def depositMoney(self, amount):
        self.__balance += amount
        print(f"{amount:d} deposited.")

    def withdrawCash(self, amount):
        if (self.__balance - amount > 0):
            self.__balance -= amount
            print(f"{amount:d} withdrawn.")
        else:
            print("Insufficient funds available.")

    def showBalance(self):
        print(f"Balance is £{self.__balance:.2f}")

#create an object from the class
newAcc = bankAccount("Mrs Jones", "12345678", 350.0)

#test our methods by depositing and withdrawing various sums
newAcc.depositMoney(100)
newAcc.showBalance()

newAcc.withdrawCash(400)
newAcc.showBalance()

newAcc.withdrawCash(100)
newAcc.showBalance()
```

The output from this test run is:

```
100 deposited.

Balance is £450.00

400 withdrawn.

Balance is £50.00

Insufficient funds available.

Balance is £50.00
```

But this is really just scratching the surface! Other powerful concepts to explore include inheritance, a powerful feature which allows the programmer to extend the functionality of existing classes, for example by creating a `savingAccount` class which adds interest calculations to our current functionality.

Whether you are programming in a modular or object-oriented fashion, it is quite likely that similar problems to those you are trying to solve have been encountered by other programmers before you. They have probably also documented their solutions somewhere, or even better they may have written a library that you can use. Of course, the question is how to find it.

DESIGN PATTERNS

It is quite common for certain types of pattern to reoccur when investigating a larger problem. Experienced software developers tend to recognise these quite quickly and are able to communicate specific design patterns to others that provide templates (or a blueprint) for solving them. Think of it as an insider's 'code'.

Although it is outside the scope of this book to cover each type in detail, it is worth noting that design patterns generally conform to one of three types:

- **structural:** increases class functionality and flexibility in usage;
- **creational:** designs for creating class instances (objects);
- **behaviour:** modifies how one class communicates with another.

Typically, these design patterns are object-oriented in nature. Although their actual implementation may vary from one programming language to another, their overall 'shape' remains constant. Typical design patterns include:

- **singleton:** ensures that only one instance of a particular class can be created;
- **decorator:** code which modifies the behaviour of an existing function or class without changing its code;
- **iterator:** creates an object whose contents can be 'looped over' (e.g. each item in a list).

And of course, there are many more.

LMGTFY! ('LET ME GOOGLE THAT FOR YOU!')

Don't reinvent the wheel. Yes, it's a cliché, but as with all clichés it contains a kernel of truth: don't create code to solve a problem if someone has already done it.

All software developers from time to time encounter a new problem, or at least something that offers them a new problem-solving experience or challenges their knowledge of their chosen language. In truth, very few problems are truly 'new', and most probably emerged in the early days of cutting-edge technologies, evolving protocols, or unfamiliar and poorly documented data formats.

Put simply, most coding problems have been seen (and solved) before, by another programmer, somewhere...

And, thanks to the internet, the solution is probably well documented (if you can find it). In truth, information for programmers has never been as plentiful and diverse, meaning that we are no longer limited to scouring manuals, readme files and Linux's manual ('man-pages').

As a general rule of thumb, software developers are encouraged to be familiar with both standard libraries (those that come with the language) and legitimate third-party libraries (those that are open source or commercially available). However, any library should be used with care, just in case it contains security vulnerabilities or limiting licensing issues. For example, a third-party library may make use of an unsecure function or its licence may preclude it being used in commercial solutions.

In addition, it is always advisable to check any issues with online community sites (Stack Overflow is a useful example). However, be warned that there's good and bad practice to be found, as techniques (and comparative efficiencies) evolve and viewpoints are, of course, highly subjective in nature. It is always worth reading:

1. the question;
2. the responses (for options and appraisal);
3. the accepted answer(s).

Other useful tips include examining solutions in other programming languages and attempting to 'port' them to your target language, although this approach can present its own challenges.

And finally, it goes without saying (but we will): don't plagiarise someone else's code!

TIPS AND THINGS TO CONSIDER WHEN WORKING WITH PROGRAMMING LANGUAGES

Here are some things that you should reflect on when you are working with a programming language. Some are tips and others are notes or suggestions.

- Don't pigeonhole yourself: many popular programming languages aren't exclusive to one particular paradigm.
- Always remember: the popularity of any given language can wax and wane over time. Don't bury your head in the sand and become too entrenched in just one technology.
- Most languages have a period of time in the spotlight; at the time of writing this is Python, whose popularity is particularly driven by its good support of data structures and third-party libraries which are ideal for data science and machine learning applications.

- Most languages have a common core of components (data types, constructs, etc.). Their actual syntax may differ, but their enabling principles are similar. This makes them easier to learn, compare and move between.
- Ensure you are aware of key concepts: arithmetic operations, algorithms and so on.
- Practise converting mathematical formulae into equivalent expressions in your chosen programming language.
- Arithmetic operators vary between different programming languages, so be familiar with the differences. For example, if you divide an integer by an integer in C, what happens?
- Practise abstraction: this is simply the concept of simplifying ideas to the essentials required to understand and solve a problem.
- Many problems involve the creation of algorithms to process data. Where possible, try to avoid O(n2), particularly when processing lists. O(n) – a linear approach – is generally better.
- Recruitment fulfils an identified need but there is a link between programming languages in active development and the available workforce with requisite skills. In other words, despite its critics, PHP is still popular as a solution because there is a deep vein of experienced PHP developers available within the workforce.
- Learn to be as self-reliant as possible in terms of research; due to the internet, information about programming languages and their techniques has never been better documented – most answers are simply a search engine request away.

SUMMARY

In this chapter, we have provided an overview of programming languages, their typical components, their prevailing trends in terms of usage and their overall market penetration. We have also touched upon the role of mathematics in software development. Many programming concepts are rooted in this discipline. While it is not necessary to have a separate mathematics qualification (although that is clearly beneficial in specialist areas such as data science and machine learning), understanding its role and relation to calculations, algorithms and complexity is undoubtedly useful.

In the next chapter we will examine the essential activities that a developer needs to complete in the role.

6 ANALYSIS AND PLANNING

'If you think it's simple, then you have misunderstood the problem.'
– Bjarne Stroustrup

This chapter provides an overview of the activities that are essential to the complex role of software developer in analysing and planning a software development.

ANALYSING THE PROBLEM

It's an undeniable fact: you can't really solve a problem until you fully understand it.

The process of analysing a problem largely depends on whether there's an existing system that is being replaced. If it exists, the system, whether manual or automated/electronic, will usually have entered a period of decay – the point at which it no longer works optimally (e.g. it can't cope with demand) – or it will no longer meet business requirements (it can't handle new opportunities).

Start by analysing the current system, asking a few key questions:

- What does the system currently do?
- What functionality needs to be improved?
- What functionality needs to be added?

Clearly the new development should retain any parts of the current system that are tried and tested and that are working well. Parts of the system which are found to be defective or insufficient need to be either rectified or replaced. By answering these questions, you will identify what the system does **now** and what it **needs** to do. The **difference** between the two represents the amount of development work required.

Of course, there may be no system already in place. In this case, there is no template to work from. While this can be liberating from a design perspective, you will need to solve all the problems from scratch.

The next part of the analysis focuses on the functionality of the new system. What does the new system need to do that the current system, if there is one, does not? If there is no existing system, what does it need to do? What problem are you trying to resolve?

As a software developer, you have a range of tools at your disposal to help you to investigate.

Identifying sources of information

When it comes to investigating systems, there are four key techniques that you can apply to obtain information. Not all of them are appropriate in every situation.

- **Interviews with individuals and/or groups:** one of the most useful techniques, interviews provide a real opportunity to explore the current system with the help of the individuals who use it. Interviews allow you to acquire information, clarify your understanding of facts, explore ideas and gather the opinions of the user base. This in itself can be very motivating for the end users, who will feel included in the development. Interviews can be either structured or unstructured. A structured interview is based on a series of questions that have been agreed in advance and each interview is conducted using this predefined list. Unstructured interviews, however, are much more fluid and so are helpful because the discussion can naturally produce useful information that you may not have thought to ask for.
- **Observations of working practices:** this technique helps to define the processes that users undertake – for example, establishing how documents flow through the system and how users interact with them.
- **Examining existing documents:** this is an often-overlooked technique, but most organisations will have a range of documentation that you can access. You should also consider examining organisational policies as this will provide information around security protocols, business continuity and backup procedures, all of which may influence aspects of the development project. Care should always be taken to ensure that any documentation being used for this purpose is up to date and reflects current practices.
- **Questionnaires:** this is another useful technique which benefits from the fact that it can be carefully prepared in advance to ensure that the right questions are asked to capture relevant responses with an appropriate level of detail. For example, you can decide whether to ask 'Do you use this feature regularly? Yes or no?' or whether you need more information, such as how often a feature is used, because that will impact on the volume of use that the feature will need to be able to support.

Choosing the right analysis techniques

When to use which technique is a key concern. Let's consider this further.

Interviews with individuals and/or groups

The key benefit to using the structured interview technique is that the responses are easy to analyse because all respondents are asked the same questions. The disadvantage, however, is that you may miss essential facts because you did not think to ask questions about a particular aspect.

The unstructured interview allows much more opportunity to explore concepts and gather information that as an analyst you may not have thought to ask for. But, conversely, you might not get answers to questions you really needed to ask but did not think to because there was no structure to the interview. In addition, the information may be quite disparate so it can be very difficult to analyse.

The reality is that most interviews will use a combination of structured and unstructured techniques, employing predefined questions to continually refocus the discussion but not limiting the discussion.

While interviews are clearly a valuable technique, they can be time-consuming to carry out and it is likely that you will not have time to undertake many. For instance, in the context of the stock system example mentioned in Chapter 5, let's imagine that the system will be used by 1,000 employees across 50 retail outlets. Although the analyst might well set up some interviews with representatives of these groups, they would be unlikely to be able to interview all 1,000 proposed users of the existing (and new) system.

Observations of working practices

The key benefits of observing working practices are that:

- You may observe processes being carried out that users might not have thought to include in a description of their working activities.
- You can see the processes in action in real time.

Observing working practices has downsides, however. A user could speed up their activities because they feel they could be judged not to be doing enough. The opposite is also possible, which is that they slow down their activities dramatically for fear that they may be given more to do. They may also do something extra when they are being observed that would not normally be part of their activity. Therefore, as an analyst you may not be confident that the timings of activities are accurate. In addition, some users may become anxious before or during an observation and in some cases the reaction could be severe. For example, they could refuse to be observed or react negatively to the observation process fearing impact on their job security. This is often connected to fear of change.

For these reasons you would be well advised to use observation only where the investigation would benefit from observing a particular process or from observing the same process carried out by a number of individuals so that average processing times can be calculated.

Examining existing documents

Examining documents such as purchase orders, sales invoices, HR records and production records will provide the most useful information about the types of data that the system will need to store; however, sensitive data may be excluded from your analysis due to privacy concerns. In general, though, examining such documents will enable you to estimate the quantities of data that will need to be handled, the frequency of transactions and the number of processes that will be involved.

Questionnaires

A carefully designed questionnaire is one of the quickest and easiest ways of capturing both simple and complex information from one or even many respondents, which is why questionnaires are heavily used in the investigative phase of software development projects. If the questions are well written, they will enable developers to capture both facts and user opinion.

However, as an investigative technique in the context of software development, questionnaires also have disadvantages. For example, many of us dislike questionnaires simply because we find we are faced with forms to complete on a regular basis (financial forms, job applications, medical questionnaires, insurance documents, security questionnaires, passport applications etc.).

Additionally, badly written questions may result in responses which inappropriately skew the results, leading to poor development decisions being made. For example, it is possible for questions to be written in a way that enables analysts to generate evidence to support their own theories. Consider the following questions. Which is likely to give the most useful (and genuine) response?

A. What are the problems with the way that invoices are currently processed?

B. What is good about the way that invoices are currently processed?

C. How are invoices currently processed?

A and B are leading questions because they direct the respondent to consider the issue from a particular perspective ('problem' and 'good'). The most unbiased question is C, and this is the most likely to generate useful information.

It is generally agreed that, when questionnaires are used in an analysis or investigation, the following are important:

- Be clear what is it you want to find out.
- Choose your respondents carefully – ensure that you ask the right people.
- Choose your questions carefully – for example, do you want to gather facts or opinion?

Next, decide on the type(s) of questions you need to use:

- **Open questions**, such as 'How are our invoices currently processed?', allow respondents to produce an extended response, but there is a danger that the respondent will miss the point of the question altogether and answer a question that is not even being asked. In addition, the responses may well be difficult to analyse because they can be so varied.
- **Closed questions** are questions where the respondent chooses from a predefined list of possible answers. This produces responses which are much easier to analyse because you simply have to count the number of responses that chose a particular answer to produce a measurable result. However, the danger here is that respondents may well have had yet another answer which you did not consider.

You can also make use of the funnelling technique, where a series of increasingly penetrative questions are used to gather finer points and detail. For example:

- Question 1: Can you tell me about customer complaints? (first question, open)
- Question 2: Which type of complaint is most common?

- Question 3: Can you give me an example of this type of complaint?
- Question 4: Was this successfully resolved? (last question, closed)

Funnelling can also be employed the other way around, starting with quite restrictive, closed questions and then broadening out into more open questioning. This often occurs during in-person interviews when an interviewee is initially quite guarded and reluctant to give detailed answers; over time they relax and are happier to handle more open questions.

Once you have decided on your approach, write the questions and decide their order. Think about the flow of the questions to help respondents move through the questionnaire as easily as possible. Think about it in the same way as you would consider tab order on an electronic form, moving logically from the top to the bottom of the form with each keypress, rather than having to jump around as the tab key moves you erratically up and down the form.

Overall, questionnaires are a very useful and widely employed investigative tool. However, their key flaws are:

- The return rate may be low as respondents who are not particularly interested may well ignore the questionnaire.
- Respondents may lie (particularly if the honest response could cast the respondent in a negative light).
- You may have to deal with irrelevant information that respondents have chosen to include.
- Respondents may not fully answer the question or may simply refuse to answer.

Putting the results of your analysis together

Once you have analysed the data gathered from the investigation, it will focus the development project, enabling the development team to identify the functional requirements of a new (or adapted) system.

You should be aware that if there is no existing documentation, or there are no users to interview or question, identifying the functional requirements can be challenging. Having said that, this situation is usually only the case for a business start-up. Any organisation which is already operating, even if it does not use a computerised solution, will still have documentation to explore and users to talk to.

Establishing requirements

The next stage is to produce a document which sets out the context of the new system and what it will be designed to achieve. This document should list the key components and processes of the existing system which will need to be incorporated into the new system. Key components can include the following.

Purpose

As suggested earlier, at the top level, the purpose of a development project will be one or more of the following:

- resolve a problem or problems with an existing system;
- respond to the changing needs of the organisation;
- respond to business drivers such as reducing costs or improving efficiency;
- take advantage of the availability of new technologies;
- respond to the activities of competitors;
- respond to the identification of emerging markets.

The purpose of the project should be written down in sufficient detail to guide the new system's designers in relation to the functional needs of the new system. The documentation should consider:

- availability of data;
- types of transaction needed;
- legal and regulatory requirements;
- administration (such as user access);
- audit tracking;
- reporting needs.

In every software development project there are also a number of non-functional requirements. These set out how the system should work or behave and may often include:

- security;
- maintenance of data integrity;
- interoperability (the compatibility of the new system with other existing components);
- scalability;
- performance (issues such as response times, volumes of data to be managed);
- reliability.

The benefit of establishing requirements at the outset is that once these are known, the development team will be able to use them to measure the success of the development project by establishing how well each of these requirements was (or was not) met.

Users

As most systems involve a level of user interaction, they should be developed to accommodate the needs of a wide range of end users. Preparing to meet user needs is something that should be thoroughly explored during the analysis phase of the development project, as understanding the user experience will ultimately generate user satisfaction. A number of factors should be considered, including aesthetics, performance, ergonomics, accessibility, usability, ease of navigation and even how enjoyable the user considers the experience to be.

Even the quality of user prompts will influence the user's judgements about the system. For example, consider the following prompt:

Input Numbur:

This simple prompt has a significant issue: the misspelling of 'number' could be interpreted by users as sloppiness and a lack of care on the part of the development team – someone should have proofread the user interface, and the error doesn't instil any confidence. Secondly and more importantly, even if it had been spelled correctly, this particular prompt gives the user no indication of what is expected. What sort of number? A whole number? Can it have a decimal part? A number in what range? This could have been improved by simply including more information:

Input a whole number between 1 and 20:

Systems programmed using visual techniques allow developers to make use of components which reduce the amount of user input, such as combo boxes and dropdown lists. This immediately reduces the potential for problems with data integrity because the user is not able to input erroneous data.

Where user input is needed, validation plays a key role in ensuring that users interact correctly with the system by rejecting inappropriate or incorrect inputs and providing guidance on what is expected. In addition, any developments which focus on **external** interactions with users, whether established businesses or start-ups, must present a good impression to users because failure to do so could mean that customers may not return to the site. Equally, a good impression could drive the business forward.

Inputs, processes and outputs

When it comes to the data component of a system, some development teams almost work in reverse by considering what the system will need to output so that the inputs and processes needed to produce those outputs can be identified. They may also consider the reports that the software will need to generate. For example:

- sales (daily, weekly, monthly, annual);
- stock (fast- or slow-moving stock, stock levels, reorder data);
- financial (income, expenditure, borrowing);
- personnel (hours worked, payroll breakdown, sickness statistics, holiday usage);
- recruitment (advertisements placed, responses).

The types of input and output really depend on the purpose of the software development project. Is it to create an electronic version of something that already exists as a paper-based system (such as an electronic rostering system for hospital staff)? If so, both the inputs and the outputs will be immediately identifiable. Alternatively, if the development project is expected to produce a completely new product, the inputs, outputs and processes may be less well known and may need to be explored.

The inputs to a system may require user interaction or may draw on historical data owned by the organisation. Alternatively, the data may need to be acquired from a third party.

Processing will include simple calculations such as invoice totals and VAT calculations, and will also include the need for data to be read from a file or written back to a file (e.g. before an item can be sold, the system may need to read the amount of stock and then reduce the stock value by the amount sold). However, processing may also require complex analytics, trend analysis, usage statistics and so on.

As a developer, you need clarity around inputs, processes and outputs as these are the heart of any system.

ANTICIPATING POSSIBLE ERRORS AND ISSUES, AND MITIGATING THEM

From the perspective of managing a development, there are a number of possible issues which could emerge. Here are a few of the more common examples.

Deadlines

In Chapter 11 we suggest that you should not 'over-promise' and 'under-deliver'. Doing so is guaranteed to create problems with your client and potentially your users, and it will often occur because unrealistic deadlines were promised at the outset. With experience, you will become better able to estimate the activities that will be needed in each phase and how long each activity will take to complete, but even so it is always a good idea to build an element of redundancy into your plan to ensure that there will be time to manage any unexpected events.

Scope creep

The requirements of the new system will have been defined and agreed at the start of the development, but it is quite common for clients to ask for changes to what was agreed, or they might even request additional features or functionality be included. This is known as scope creep (as already touched upon in Chapter 4).

Avoiding scope creep can be a challenge, particularly if you want to keep your client happy. But responding positively to these requests can drive up the costs of the development due to the time taken to accommodate them. For this reason, smart development teams will often estimate the cost of implementing the request and will raise an invoice for that value which they will share with the client. In reality they may or may not choose to raise an invoice, but producing the invoice will show the client the financial impact of such requests and will make them think more carefully, particularly when they understand that accommodating such changes could mean having to pay more for the development project.

Having said all this, some of the methodologies discussed in Chapter 4, particularly Agile, are based on the notion of the product emerging over time and therefore scope creep is part of the process. As a developer, you will need to monitor these requests and ensure they do not become excessive, entail significant additional cost for the client, or morph the development into something completely different from what was intended.

When it comes to scope creep, more rigid methodologies, such as Waterfall (see Chapter 4), can help you to maintain more control.

Poor team communication

It is now very common for development teams to increase their coding capacity by subcontracting aspects of development to third parties, often overseas. Part of the reason for this is to access a wider talent pool while keeping costs low, because the average hourly rate of programmers in countries such as India and Poland is far lower than that in countries such as the UK.

This can result in communication issues, particularly in terms of ensuring that the subcontractors fully understand what it is you are asking them to do. It is therefore essential to closely monitor the activities of the team to ensure that the development project is progressing as expected. This should be done regularly using a variety of methods, such as code reviews, daily stand-ups, stakeholder briefings and project management platforms, so that any issues can be quickly uncovered and dealt with.

Poor team collaboration

Poor collaboration can result from a lack of, or poor, communication. In addition, your development team could suffer from a variety of issues which could affect their successful performance. For example:

- The size of a team can affect how well its members work together because with a large team you simply have a larger number of personalities.
- Larger teams may need to be divided into sub-teams. This allows each sub-team to be assigned specific tasks and their concerns, input, questions and so on to be communicated via a sub-team leader, effectively streamlining discussions and keeping development meetings suitably focused.
- The team members may not all be prepared to share information.
- Pre-existing conflict can resurface, particularly during times when development activity is stressful. This can be a particular issue if you have domineering personalities in the group.
- A lack of clarity about how the team will be rewarded for their activities can result in a sense of unfairness.
- A lack of focus can result if team members are not on board with the overall objectives and ultimate goal.
- When team members are based in different countries, there can be both physical and cultural barriers which need to be carefully managed.
- There may not be a feeling of mutual trust across the whole team.

Potential ways of promoting collaboration include:

- Promoting a positive attitude among team members is essential to effectively managing a team.
- To avoid individual team member disengagement, all team members should have their ideas heard.
- Promote a positive team culture and avoid the blame game. If a member of the team has problems, the team should pull together to help resolve the situation rather than simply blaming the individual personally for their failure.

Access to staff

One of the most irritating problems when you have put together a development team is when one or more members become unavailable. Holidays, for example, can be factored into the planning, but sickness can usually not be predicted.

There can also be situations where team members are pulled off a project to be assigned to other activities. This in particular destabilises the team when it creates a skills shortage. Finding a replacement can be difficult and, even if you do, you then have the additional problem of bringing the new team member up to speed.

See more on this topic in Chapter 4.

Lack of staff understanding of their own role and responsibilities

Whether the entire team is involved in the decision-making and planning processes or not, there should be clear demarcation of roles and responsibilities. Staff should be supported to ensure that they understand their roles within the team and within the context of the development project. If this does not happen, team members could disengage with the process and put the entire project at risk.

Not being aware of an impending problem

Even if you have taken steps to mitigate possible problems, there are not many development projects that do not suffer from some sort of issue from time to time. To give yourself the best chance of identifying emerging issues and addressing them, make sure to carefully observe team activities and listen carefully to the concerns of both the team and the client or users.

Ideally a development should aim to 'design out' any potential operational problems that could occur later. This means proactively reworking ideas so problems don't occur and therefore **do not need to be solved**. For example, changing a user input from free text to a selection box of available choices removes many validation-oriented issues.

Designing systems that will be robust is usually achievable if you fully understand your data and the needs of your client and users. But there are other principles that should be considered too, such as scalability. It is very easy to make a product which meets the immediate needs of the client. However, the client's needs may outgrow the original design during the development period. This could be because the system has not been designed to handle a larger number of future transaction or data volumes, cannot cope with an increased number of users, does not have sufficient storage availability, or generally is unable to handle an increased workload as time passes.

THE ROLE OF DOCUMENTATION

During planning, consideration is given to documentation that can be generated during or at the end of the development. All decisions made and the development activities undertaken are recorded to provide data that will be used in the review. This ensures that lessons can be learned.

The documentation that is typically generated includes:

- development plans and schedules;
- group activity logs with expected activity dates – the logs are then annotated with monitoring detail to record what is complete, what went well or what went badly, and information about changes that had to be made or how emerging problems were solved;
- diagrams representing both the existing system and the design of the proposed system, such as flowcharts, data flow diagrams and system diagrams;
- annotated code;
- test data;
- client feedback received throughout the development project – the plan may be that all feedback will be provided by the client via email (rather than verbal feedback), so that the email or a sequence of emails can provide a permanent record;
- minutes of development meetings (if these occurred);
- photographs of Scrum boards (or archived digital copies) – during planning, decisions will be made about whether physical Scrum boards will be photographed, or whether the Scrum boards will be digitally documented using proprietary software such as Asana, Wrike or Workamaji®, with the advantage that the Scrum boards can also be accessed on devices as the data will be held in the cloud.

Some of the documentation listed above will ultimately be included in the technical guide to support the maintenance of the product. These documents will provide a focus for discussion and enable team members to learn lessons from the development activity.

SUMMARY

In this chapter we have reviewed the essential activities in relation to analysis and planning with which a software developer might engage as part of their role. Perhaps the most important point to remember is that it doesn't matter how good you are at creating software if you don't truly understand the underlying problem you are trying to solve.

In the next chapter, we will explore how to write good-quality code.

7 WRITING GOOD-QUALITY CODE

> 'Programs must be written for people to read, and only incidentally for machines to execute.'
>
> – Harold Abelson

It is not enough to solve a problem if the resulting code is not maintainable. Does it really matter if the code is inefficient, poorly laid out or not commented? In this chapter we consider the answers to this question.

CODING THE SOLUTION

Although it is outside the remit of this book to provide detailed instruction in any given programming language or recommend which one to use in any particular circumstance, there are some techniques you should consider no matter which operating system, hardware or language you are using.

In an ideal world

The heart of software development activity is the production of working code, progressing through the following stages: entry of source code, building of executable code (where appropriate), testing, staging (where relevant) and deployment.

Ideally any solution should:

- produce accurate results
- and do so efficiently
- while working reliably
- and robustly
- while also being coded to appropriate standards
- and being secure
- and most importantly meeting the design requirements
- within the budgeted time allocation.

Of course, this is not an easy set of criteria to meet, and modern software often stumbles in one or two of these categories, often in terms of reliability and security. This is not surprising perhaps, as many organisations do not allocate sufficient resources to test program code thoroughly. It's an unfortunate shortcut, but one which is often accepted as the price of getting solutions 'live'.

Make no mistake, though, it can be an exceptionally costly and short-sighted decision. There are often unforeseen consequences that karmically end up costing an organisation even more money than implementing a decently robust testing regime. It's a tragic and false economy.

Unfortunately, as programmed solutions infiltrate more and more of everyday life, their need to be as bulletproof as possible will only grow more critical. This software will determine medication dosages for critically ill patients, decide how to drive an automated car on public roads and monitor the safety levels in nuclear power stations. The consequences of less-than-perfect reliability and security cannot be underestimated.

How coding is achieved commercially

Producing code often comes down to a personal choice of available editors, compilers, debuggers and operating systems. These create a 'toolchain' which enables the developer to create, execute and test their code to the required standard.

In a commercial environment, you will often find two developers sat side by side using very different environments, each singing their choice's praises and recoiling in horror at the tools chosen by their peer. Some may prefer a powerful editor such as Vi and command line compilation, whereas others may prefer the auto-completion richness of an IDE (integrated development environment) such as IBM's NetBeans or Microsoft's Visual Studio.

The list of development tools in use may be mandated by an organisation, but it is not uncommon for programmers to have some degree of latitude, excepting (of course) the actual programming language being used – that is often specified as part of the design and is not a subject for discussion. Commercially it is not uncommon for some organisations to pre-install a new developer's environment to ensure they are ready to be 100% productive from day one.

Infrastructure-rich organisations often rely on enterprise glue to fix parts of the solution together – for example, to download data files from an FTP server for processing or to automatically download and print generated invoices. In these instances, some freedom may be available, particularly in scripting languages. For example, a programmer may decide on a Linux platform to code in shell script, Perl or Python and achieve similar results. The ability for the organisation to **maintain** the code in the future may affect the decision here, particularly if expertise in certain languages is known to be thin on the ground.

Although diversification in practical workflows is therefore accepted, the rise of the Agile methodology as the premiere approach in commercial development has become a unifying driver.

The importance of version control

Being a commercial developer typically means working in a team. This will directly impact the code you write and how you contribute to a project. In a modern software development workplace, this is typically achieved via software versioning. This is examined in greater detail in Chapter 14.

NAMING CONVENTIONS

Giving identifiers suitable and meaningful names is only part of the equation. The other aspect to consider is the naming convention(s) being used in a software development project. The four common styles are shown in Table 7.1

Table 7.1 Naming convention examples

	PascalCase or TitleCase	camelCase	snake_case	kebab-case
Example	GrossPay	grossPay	gross_pay	gross-pay

Naming conventions are typically either mandated by the organisation (as an internal coding standing) or by the programming language itself. In an organisation, naming conventions simply make life easier for all developers when a project (and its team) is large; in theory, with a consistent naming convention, any developer should quickly be able to read the code produced by another.

In languages such as Python, PascalCase (following that language's popular usage of the styling), camelCase (due to the 'hump') and snake_case (lying low to the 'ground') can be used to indicate that an identifier represents a class, a function or a variable, respectively. However, organisational standards and those shared by a development team will often prevail.

For the record, if you're wondering where you might encounter kebab-case, it's most often used by web application developers when they create CSS classes to format their HTML-based content.

THE IMPORTANCE OF LAYOUT AND COMMENTING

Layout typically takes the form of indentation, the process of highlighting the logical structure of program code by indenting certain parts of code from the left margin. An `if` statement is a typical example of this. In C, this would appear as follows:

```
if (age > 18) {
  printf("You are older than 18!");
}
```

And in Python it would appear as follows:

```
if age > 18:
    print("You are older than 18!")
```

Although the syntax is slightly different, both languages use indentation to highlight that the printing of the message is connected to the evaluation of the expression. In other words, if the condition is true (age is greater than 18) then (**and only then**) will the indented code be executed.

It is worth noting that in C the indentation is purely aesthetic and has no impact on the logic; C typically uses braces to create blocks of code, which it connects to controlling expressions. In Python, however, the indentation is an important syntax point: it is mandated as four spaces (*not* a tab) in PEP8 (Python Enhancement Proposal 8), which is **the** style guide for all Python developers.

COMMENTS AS DOCUMENTATION

Documentation is a significant part of the development process. For some programmers it comes naturally as they start writing comments from their initial pseudocode as an outline of the steps they intend to take to solve the given problem. Pseudocode is a step-by-step list of actions that the expected program will need, written in the user's native language. You will find more on pseudocode in Chapter 11.

For example:

```
get first number
get second number
result <- first number + second number
output result
```

However, it is often derided in industry through the oft-quoted conceit of making program code 'self-documenting'. This idea suggests that if the code is clearly written, uses meaningful identifier names and has good indentation then it shouldn't need separate documentation – any reasonably competent programmer should be able to understand it 'as is'.

That's true to an extent but it doesn't tell you the whole story.

The one thing you shouldn't do when commenting is to just describe the actual syntax being used, as in this example in C:

```
int qty;
//set quantity to 0
qty = 0;
```

It's obvious what's happening here, so adding this all-too-obvious comment is essentially a waste of time, effort and an employer's development budget! This is sometimes referred to as WET (write everything twice or, more humorously, we enjoy typing), the opposite of the software engineering principle DRY (don't repeat yourself!).

Of course, many programming languages support documentation features and although there are differences in their syntax, the basic objectives of good commenting are clear:

- Concisely **describe the actions of your code** to others **from a real-world perspective** (what it does, **not** how it does it).
- Provide specific **programming notes** – for example, why something is included that might be platform specific (necessary for Microsoft Windows or Linux, for instance) – that might be useful and not immediately obvious.
- Provide an **overview to other programmers** who must maintain your code so that they can glean its purpose after a quick skim read.
- Act as an **aide-memoire** when you return to revise the very same code in the future.

The following Python code extract is a useful function which converts a temperature in degrees Celsius to its equivalent value in Fahrenheit:

```
def c_to_f(celsius: float) -> float:
    """
    A function which converts a temperature in Celsius
    and converts it to its Fahrenheit equivalent.
    """

    fahrenheit = (celsius * 9/5) + 32

    return fahrenheit
```

These comments (and hints) tell us:

- the type of the reference (data) passed into the function (a floating-point number);
- the type of reference returned by the function (another floating-point number);
- the function's purpose (stored as a Python docstring).

And, more helpfully, after importing its parent module, the new function is immediately usable by Python's built-in help function:

```
>>> help(c_to_f)
Help on function c_to_f in module __main__:

c_to_f(celsius: float) -> float
    A function which converts a temperature in Celsius
    and converts it to its Fahrenheit equivalent.
```

In addition, Python's type hints (specifying the data types passed in and returned by the function) can be accessed by various IDEs, offering developers a helping hand as they type:

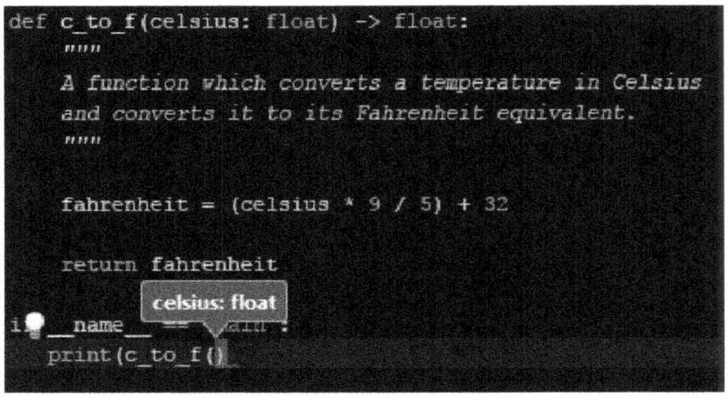

```
def c_to_f(celsius: float) -> float:
    """
    A function which converts a temperature in Celsius
    and converts it to its Fahrenheit equivalent.
    """

    fahrenheit = (celsius * 9 / 5) + 32

    return fahrenheit
                celsius: float
i  __name__
    print(c_to_f()
```

Using documentation generator tools

Many tools are available which will generate HTML files that can be viewed through a standard web browser. They typically offer an interactive experience for a programmer, enabling them to easily navigate a hyperlinked version of their complete software project function by function, class by class and so on.

Some tools available for the more popular languages include (the clues are in the names):

- pydoc (part of the standard Python library; see https://docs.python.org/2/library/pydoc.html);
- PhpDocumentator (www.phpdoc.org);
- Javadoc (part of Oracle's Java Development Kit installation; see https://tinyurl.com/ydhgcray).

For example, issuing the following command in a Microsoft Windows environment:

will generate HTML documentation for the previously discussed Python **`c_to_f`** function, which exists in the **`degcf`** module (see Figure 7.1).

Figure 7.1 Automatically generated HTML documentation for a Python function

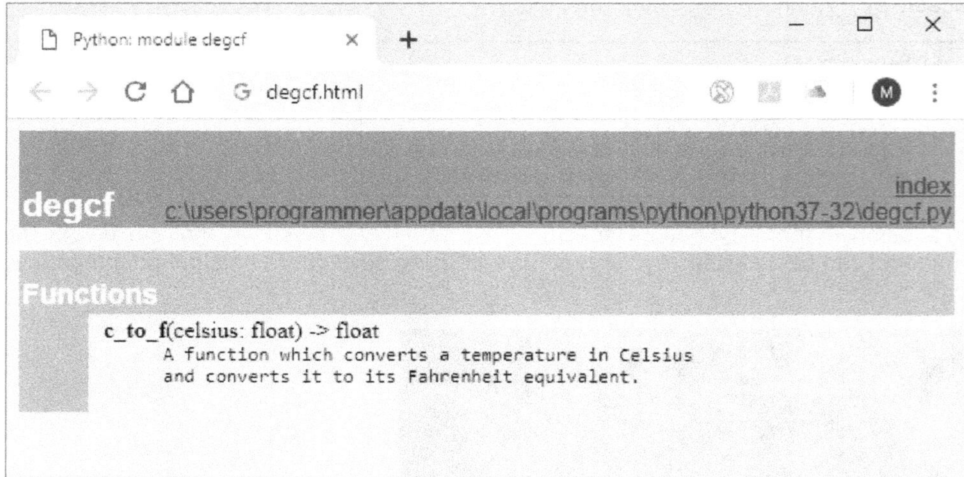

Again, use of Python's comments has proven very helpful, producing a simple but effective overview of the function we have written.

As you may imagine, a larger project would create a much larger and richer repository of information that any software maintainer would be happy to peruse.

HANDLING ERRORS AND EXCEPTIONS

Handling run-time errors

In many programming languages, your source code is translated into machine code (binary) for the CPU to execute using a process called 'compilation'. When this process cannot be completed wholly successfully, it results in errors and warnings:

- **Error:** a fatal error, typically a problem with the language syntax where the programmer has broken the rules of the language in their code, such as by missing out a bracket or semi-colon. These must be fixed before the program can be successfully compiled and executed.
- **Warning:** a non-fatal error, which is advisory, typically indicating an issue that should be investigated as it may cause a problem at run-time. These typically can be ignored, but the wise programmer does so at their peril!

A typical warning could be the narrowing of a data type, for instance storing a floating-point number (one with decimal places) in an integer variable (a whole number), in which case the accuracy of the original data would be lost through the truncation of the decimal part of the value. Not fixing this could generate wildly inaccurate results, and in real-world contexts it could prove costly or lead to disastrous outcomes.

However, some errors cannot be spotted during the compilation process. These errors can occur unexpectedly as a program is executing; these are called 'run-time' errors. In practice, a programmer should code defensively against these, anticipating the problem and providing a suitable solution which handles them; this leads us to exception handling.

Handling exceptions

When a computer program runs some operations, unexpected events can occur, for example an attempt to divide by zero (which results in infinity, something a computer system cannot store). Under normal circumstances, these serious events would cause a fatal run-time error, typically crashing the program (or worse, the entire operating system), necessitating a reboot.

However, in many modern programming languages it is possible to 'throw' an exception in response to such errors. Exception handling is then used to manage this type of problem, effectively 'catching' the exception and 'handling' it with suitable code, allowing the program to continue unhindered.

This is typically achieved using **TRY...CATCH** blocks:

```
TRY
  <statement>
CATCH (exception)
  <statement>
```

Many different types of exception can occur, each requiring a specific handler. Of course, although the actual syntax for these exception-handling blocks varies between languages, their structure follows a familiar pattern.

Worked example

A common form of run-time error occurs when a program attempts to open a data file which cannot be found, perhaps because it has been moved, renamed or deleted. Here is a Python 3.X excerpt which demonstrates this:

```
error_filename = "file_import.log"
error_file = open(error_filename, "a")

#...processed filename, e.g. from an FTP server
filename = "100420.dat"

try:
    file_to_process = open(filename, "r")

except FileNotFoundError:
    print(f"{filename:s} not found", file=error_file)

else:
    for line in file_to_process:
        print(line, end="")
    file_to_process.close()

finally:
   error_file.close()
```

In this example, a **FileNotFoundError** exception is thrown at run-time when the computer attempts to open the named file. This exception is then caught and gracefully handled by appending a simple message to a log file, which the developer can later examine. Notice that we only attempt to read the file if it is successfully opened; this 'successful' action is indicated by using the **else** clause. The **finally** clause is executed regardless of the successful opening of the data file. This prevents a serious run-time error in the program.

As discussed, another common form of run-time error occurs when a program attempts to divide by zero, something which is arithmetically impossible. The following **TRY...CATCH** style of exception handling prevents this type of error occurring in Python 3.X:

```
total_sales = float(input("Enter total sales: "))
days = int(input("Number of days: "))
try:
    sales_per_day = total_sales / days
except ZeroDivisionError:
    print("Must enter 1 or more days.")
else:
    print(f"Average sales per day: {sales_per_day:.2f}")
```

In this example, a programmer is trying to calculate the mean average sales over a number of days. However, because the number of days being input has not been validated to check that it is a positive integer, it is possible that a division by 0 could be attempted.

A simple **TRY...CATCH** block around the calculation will avoid this problem and handle the resulting **ZeroDivisionError** exception with a friendly error message. A better solution would be to validate the user input, of course – another example of designing the problem out so you don't have to deal with it.

The exception handler is just a specific block of program code that deals with thrown exceptions. Typically, there are multiple handlers, one for each type of common exception caught by the developer. This is **much better practice** than having a generic handler that catches everything but cannot act specifically to resolve them.

Why exception handling is important

There are a number of reasons why exception handling is important:

- It is not possible to anticipate all errors that might occur at run-time when designing a program.
- It separates error-handling code from normal program code, which eases code maintenance.
- It is capable of recognising and handling different types of error – for example, a file not existing raises a different exception to a file being of the wrong type.
- It prevents run-time errors from causing programs to fatally crash or become non-responsive.
- The successful handling of run-time errors is particularly important for automated scripts which run outside normal office hours (such as overnight file downloads) as if they spectacularly crash, no developer will be on-site to rectify the problem.

Exception handling is a crucial element in many commercially important programming languages in the IT sector, including Python, Oracle's Java, Microsoft's C# and Visual Basic .NET, Apple's Objective C and Swift.

Programming language idioms

All programming languages have their own unique flavour, typically represented by the ways certain tasks are commonly performed. These language-based 'code patterns' are traditionally used when performing certain algorithmic tasks, and they shouldn't really be translated from one language to another without due consideration.

For example, in C, it is a tried-and-tested tradition to iterate along an array using a `for` loop and an index:

```
int data[5] = {10, 20, 30, 40, 50};
for (int i=0; i < 5; i++) {
  printf("pos %d value %d\n", i, data[i]);
}
```

This C algorithm can be mechanically replicated in Python:

```
data = [10, 20, 30, 40, 50]
i = 0
while i < 5:
    print(f"pos {i} value {data[i]}")
    i += 1
```

However, most professional Python developers would not code in this way, as it does not use a **recognisable** idiom. Either of the following patterns would be considered far more 'Pythonic':

```
data = [10, 20, 30, 40, 50]

for i in range(0, 5):
    print(f"pos {i} value {data[i]}")

for pos, value in enumerate(data):
    print(f"pos {pos} value {value}")
```

It is interesting to note that even the use of `i` as an index in a loop could be considered a universal idiom – it's a nugget of knowledge that all developers inherently understand as an accepted practice and, hence, are not surprised when they see it being used in a coded solution.

Using a programming language's idioms is a way of using its features appropriately, writing in a more naturally expressive manner and, consequently, producing better code.

EAFP vs LBYL

Traditional programming follows a LBYL (look before you leap) approach to coding. For example, we check to see if a file exists **before** we try to open it. This makes a lot of sense and it avoids the embarrassment of the underlying operating system reporting a 'file not found' error.

However, an EAFP (easier to ask for forgiveness than it is for permission) approach works the opposite way around. If we use this approach, we make the basic assumption that the file (always) exists and try to open it. Placed around this attempt to open a file is a simple **TRY...CATCH** block which will handle the generated error.

Although the former may seem more sensible, it is the latter approach that is favoured in popular languages such as Python.

Refactoring

It is not a trade industry secret to say that code is often developed at a reasonably frantic rate. This leads to technical shortcuts, a lack of appropriate (i.e. rigorous) testing and, to be honest, less-than-ideal additions to a production codebase. The usual suspects include:

- vulnerable code with insufficient data or data type checking;
- overlong (or overcomplicated) functions;
- functions which break the SRP (single responsibility principle): performing more than one core task;
- badly designed classes, particularly ones that have grown too large and unwieldy;
- orphan functions which possibly should belong elsewhere;
- unnecessary code duplication: WET vs DRY;
- short-term 'hacks' which are used to patch a codebase's behaviour temporarily;
- obsolete and/or potentially misleading comments which no longer reflect the code they accompany;
- over-engineered code which attempts to predict future need (more on this later) and becomes needlessly layered and overcomplicated;
- badly named identifiers – too short, too long or simply do not meet team standards;
- 'God' lines, where a single complex line of code possibly defies easy maintenance or breaks coding standards (e.g. recommended maximum line length).

When tight release deadlines loom and code 'just works', it is common (and human) for hard-pressed developers to simply walk away and close the box, often hoping that any further revision to the codebase will be assigned to another developer. When this happens, the team (and the project itself) is effectively carrying 'technical debt'. The longer this debt is carried, much as in real life, the more 'interest' will accrue. One solution is to consider refactoring.

Refactoring is a process which, ideally, shouldn't stand out as a separate (or special) activity in a software development team's workload. It involves reorganising existing code while not changing its ultimate behaviour. Often, in the first gasp for air after a project has been crunched to completion, time is found to perform some refactoring, particularly if the code 'smells'. Code smell is a purely subjective concept; it is essentially defined as the innate feeling that something in the codebase 'isn't quite right' and may be the cause of deeper problems (i.e. in performance, accuracy or reliability).

In truth, it is generally better to perform refactoring while new code is being written (i.e. as part of a developer's everyday routine). In other words, when working on a sub-task, if you notice some existing code that smells, you can simply opt to refactor it there and then. This contributes to the concept of a team's sense of collective responsibility over their codebase – in itself, no bad thing.

Selected examples of code smells and their refactoring solutions are shown in Table 7.2.

Table 7.2 Refactoring solutions for a range of code smells

Code smell	Typical refactor solution
Vulnerable code with insufficient data or data type checking	Add extra 'belt and braces' checks for data existence, data range and data types (assume **nothing**)
Functions which break the SRP (performing more than one core task)	Break the larger function into a number of smaller functions which each meet the SRP – this is by far the most common refactoring activity
Unnecessary code duplication (WET vs DRY)	Identify the duplicated aspects and create suitable functions
Orphaned functions which possibly should belong elsewhere	Consider placing each function in a library or add it to an associated class as a static method
Short-term 'hacks' which have been used to patch a codebase's behaviour **temporarily**	Re-open the box and solve the issue **permanently** – hacks should **always** be removed!

Many modern IDEs, such as Eclipse and NetBeans, include automated refactoring routines which can assist a developer in the refactoring process.

It is critical to remember that refactoring should change the internal composition of the code but **not** affect its external behaviour – certainly not in a negative way. If the resulting code executes faster, is more efficient in its memory usage or simply behaves in a more robust manner, complaints are unlikely to be forthcoming.

The exact working practices inside a software development team can vary between organisations so don't expect them all to work in the same manner. Agile processes entail many different aspects, such as pair programming (see below). When implemented effectively, this can certainly make a positive contribution to the writing of good-quality code.

CODE REVIEW METHODS

Pair programming

Pair programming is a common and generally productive Agile software development technique which involves two programmers working together on a single problem at the same terminal (Figure 7.2). The programmers generally have similar levels of experience (this isn't a traditional master–apprentice arrangement). The arrangement provides an opportunity for each to take turns at coding and reviewing, switching roles regularly and most commonly during changes of module (or function).

Figure 7.2 Pair programming (This photo by Unknown Author is licensed under CC BY-SA-NC)

While one programmer works on the current code, the other reviews and makes suggestions as the code develops line by line on the screen, often identifying potential improvements to the code or correcting syntax or semantic errors as each block of code is written. Discussion and compromise form part of the process when viewpoints don't coincide; this is natural.

The impacts of pair programming techniques are mixed, as shown in Table 7.3.

The biggest positive impact of using peer coding techniques is that they are very true to the Agile philosophy: feedback on potential mistakes, logic flaws and so on are kept as close as possible to the point of the code's creation – even before the code reaches the quality assurance process and, better still, before it's in the hands of the actual customers. In this way, and with working software being the key measure of progress in any Agile environment, peer coding practices work well and should always be considered.

Formal peer code reviews

An Agile development environment offers another mechanism that can encourage the creation of good-quality code – the formal peer code review. In a modern development

Table 7.3 Pros and cons of pair programming

Advantages	Disadvantages
• Programmers tend to enjoy the experience and it can increase confidence in a solution working as expected. • It improves communication, negotiation and other common social skills, which can aid more introverted programmers. • Knowledge is shared; programmers learn new tactics and techniques from their peers (this can act as a progressive form of continuing professional development). • Different ideas are considered during the coding process, permitting filtering of better ideas and discarding of weaker ones. • Different viewpoints tend to consider broader areas of impact of new code on the larger system. • The produced code tends to have fewer defects (especially run-time errors caused by semantic errors, as these are often caught much earlier by the peer); this can offset developer costs longer term. • Solutions may be developed faster. • The process can complement lightweight code reviews. • The process can occur remotely using real-time editors on a shared desktop.	• It typically increases the number of developer-hours needed to solve a problem... • ...which can noticeably increase project costs! • It can be difficult to (initially) convince management of its effectiveness as an implementation strategy. • It requires both programmers to contribute evenly to the process. • It does not work well if the programmers do not have similar levels of technical skill. If the programmers are very mismatched, one will tend to withdraw from the process and not question the code being produced. This can also happen when the programmers are both very experienced as neither may feel the need to question the conventional wisdom of solving a problem a certain way. • It is not appropriate for simple tasks that could be solved by a single programmer; over-engineering may occur.

environment, all members of the team should be encouraged to take a collective attitude to maintaining the underlying state and quality of the active project's codebase. There are many reasons for this:

- Developers gain a wider understanding of the entire codebase, offering more redundancy in the team (i.e. more ability to cope with illness, members leaving etc.).
- Common coding standards are easier to communicate and adhere to, giving the codebase a more uniform appearance and problem-solving approach (this also benefits new programmers as the style is immediately easier for them to spot and adopt when they start).
- Code documentation tends to improve.

It is generally cheaper to fix identified bugs earlier in the development process rather than later. Early code reviews, achieved through the collaboration of a small team of reviewers, encourage the elimination of shoddy programming practice, use of ill-considered solutions and unrequested deviations from the agreed design documentation.

Pair programming can, of course, be used as a lightweight method of peer code review. In addition, panel reviews via online tools or simply email-based message threads can be effective. It is also common to select a subset of new sections of code to review rather than to review literally everything. The code that is selected may include those sections that cause the most concern among the team (potentially from a new programmer) or those which contain the most critical (and therefore impactful) additions to the existing codebase.

How could the usefulness of code reviews be evaluated? We could simply analyse the relationships between the following variables to generate a useful metric:

$$\textit{Review impact} = \frac{\textit{Number of identified issues}}{\textit{Lines of code}} \times 100$$

In addition, the actual time (and associated costs) required to review the selected code should be factored into any calculation.

As with all review techniques, there can be resistance, especially if a development team does not buy into the process. They (and you, assuming you're part of the team) simply need to remember that the goal is to measure the quality of the produced code, not make judgements about the ability of its author: we're all capable of producing poor-quality code on any given day – it's just a part of being human.

TIPS FOR GOOD CODING

Here are some things to remember to help you to produce good code:

- Use meaningful names when creating identifiers (i.e. variables, constants, classes, functions, structures etc.).
- Ensure each function has one purpose, according to the SRP.
- Delete any unnecessary code – if you're using version control it'll still be safe in the change log.
- Readability **always** counts.
- Code in a consistent manner and follow organisational standards.
- Indent your code appropriately.
- Comment the purpose of your code, not the syntax.
- Never add comments which could be seen as commercially insensitive, rude or unprofessional – you never know who will see them.
- Use the idioms of the language whenever possible.

- Always catch exceptions and never let them pass without comment – logging them is a good start.
- Refactor where appropriate and do it proactively rather than reactively.
- Pair programming can reduce errors and improve the quality of the code produced.
- Play an active part in formal code reviews. There's always much to learn no matter how experienced you are!

SUMMARY

In this chapter we have examined how to write good-quality code. In truth, it is difficult to get it 100% correct, as readily evidenced by the proliferation of bugs, patches and broken applications you have no doubt encountered in your travels. However, of course, that doesn't mean we shouldn't try.

In the next chapter we will turn our attention to the creation of effective user interfaces.

8 DEVELOPING EFFECTIVE USER INTERFACES

'As far as the customer is concerned, the interface is the product.'

– Jef Raskin

'If you think good design is expensive, you should look at the cost of bad design.'

– Ralf Speth

Even if the code is efficient, is meticulously executed and has all the required functionality, a poorly designed and developed user interface will lead to a poor user experience and ultimately a lack of confidence in the product.

In this chapter we will explore the concept of user interfaces and how these can affect the user experience.

USER INTERFACE AND USER EXPERIENCE

Left to our own devices, many of us will tend to create a user interface (UI) based on our own preferences. While that would obviously suit other users with the same preferences, it will probably not suit the majority. This applies not just to the UI but also to the whole user experience (UX) with a software development product.

The term 'user interface' refers to the technologies (screens as well as input and output peripherals) that allow the user to interact with the computer or a device. The term 'user experience' is the level of enjoyment or frustration a user faces when interacting with the computer or device.

Here are some general things to consider when working on the UI and UX components of a software project:

- Investigate the preferences and requirements of users, drawing on a wide range of potential users of the software. This will ensure that you know your audience.
- Understand how your users will interact with the interface.
- There will be times when you will need to make assumptions. You should ensure that any assumptions you make are validated with users across the range of user groups.
- Ensure that UX is high on the list of considerations. An appreciation of the UX is as important as the functionality. A poorly designed UI will lead to a poor UX and may lead to a reduction in users even if the software functionality is outstanding.
- Avoid any potential for poor technical experience – this is particularly relevant with the development of web apps, which may need to function across a range of browsers.

- Ensure that your software is clear and simple to navigate and that any help is easy to access.
- Ensure the consistent use of fonts. Find different ways to emphasise aspects of your interface rather than simply resorting to changing the font.
- Make sure that error messages are useful and give users a clear indication of what they should do next in the event of an error.
- Manage user inputs. Give your users visual guidance on the type of input expected (e.g. 'Enter height in metres (e.g. 1.3)' or 'Enter height in centimetres (e.g. 130)'. Simply prompting a user with 'Enter height' could elicit either response.
- Help users to input data correctly by implementing validation to limit input and make sure that a user's inputs are at least sensible.
- Reduce the need for user input by employing mechanisms that enable users to choose from appropriate options via dropdown boxes, combo boxes, radio buttons, check boxes and so on.
- Make the UI easy to navigate and use.
- Concentrate on the alignment of prompts and input boxes, and ensure that there is white space to reduce the level of screen noise.
- Declutter the UI as much as possible (using pop-up menus or links between screens) to reduce the likelihood of overwhelming users.
- Limit the number of colours in use and choose simple colours.
- The best UI designs should be relatively intuitive.

USE OF TOOLS

Depending on the type of project, many different UI and UX tools are used. Some help with designing interfaces while others are used to create prototypes or communicate potential choices to clients.

It is beyond the remit of this book to cover all possibilities, but some common ones include:

- **Storyboards:** visual representations of each of the screens that the user will navigate. The term is taken from the film industry, where storyboards are used to plan the narrative of a film or scene.
- **Moodboards:** particularly used in web development, these show how images, styles and colours will be used together to evoke a concept or style.
- **Wireframes:** these are visual representations of the functional aspects of the UI and how they interconnect. There is no styling or branding. Wireframes are very popular in mobile application development.
- **Interactive prototypes:** these are stripped-down versions of the intended product, with limited functionality that the user can experiment with.

In addition, traditional graphic design packages can be used to create artistic mock-ups of UIs. However, modern UI and UX trends focus on providing a more interactive client experience, demonstrating how the end user engages with the software, not just what it looks like.

Key UX design principles include:

- Meet the needs of your users (simplicity for inexperienced users, but consider more complexity for experienced technical users).
- Have a clear plan. Know how one screen links to another and how users will navigate between screens and the home page or menu. Have a clear hierarchy that sets out how the content of the program, web page or app is organised.
- Ensure you have consistency – this can be in terms of the use of buttons or hyperlinks to navigate between elements, but probably not both as this can be confusing for users.
- Think carefully about accessibility and design your solution to cater for the needs of the widest possible range of users who might want to use your software. This may mean making adaptations to suit users with visual impairment, physical challenges, deafness or other restrictive conditions. For example, if you are focusing on web development, you may find it helpful to reference the World Wide Web Consortium's web content accessibility guidelines (W3C 2018).
- Don't forget about the context of your product. Business solutions should look and feel professional, whereas you can be less formal when developing more social products.
- Don't forget about usability. Solutions which are busy and cluttered with lots of buttons and links can lose users' attention very quickly.
- Consider the language you should use for your particular audience. Simple language will always appear more user friendly.

If you want to explore this topic further, you are encouraged to read *User Experience Foundations* (de Voil 2020).

SUMMARY

When software developers learn to program, they should think about their users. It can help to put yourself in the shoes of users by thinking about elements of software design that you yourself don't like. If you do not like something, why should someone else? Use the available tools and consider the suggestions outlined in this chapter.

In the next chapter we will examine how program code can be linked to a variety of back-end data sources.

9 LINKING PROGRAM CODE TO BACK-END DATA SOURCES

'Unity, not uniformity, must be our aim. We attain unity only through variety. Differences must be integrated, not annihilated, not absorbed.'

– Mary Parker Follett

Modern software solutions tend not to rely exclusively on keyboard input and screen output. Although it's traditional for new developers to cut their teeth on console-based applications which accept keyed values, process them and then output to a screen, this is quite atypical in the commercial sector, highly instructive as it may be! Current solutions are often based on the integration of different technologies, working together.

This chapter focuses on different data sources and demonstrates how they can be incorporated into a working solution.

SOURCES OF DATA

In the IT industry, developers draw data from several sources, including but not limited to:

- Hardware interfaces, such as sensors.
- Data files:
  - unstructured text files;
  - structured text files, such as:
    - Comma-Separated Values (CSV);
    - Extensible Mark-up Language (XML);
    - JavaScript Object Notation (JSON);
    - YAML Ain't Mark-up Language (YAML);
  - binary files.
- Databases:
  - SQL (Structured Query Language) databases;
  - NoSQL (Not only SQL) databases.
- Web-based APIs.

HARDWARE INTERFACES, SUCH AS SENSORS

The use of sensor technologies has dramatically expanded over recent years with the benefit that equipment and environments can be monitored and automated or managed manually from remote locations. Computers have a number of sensors which monitor various hardware components and their status, such as fan speed and temperature.

Gyroscopes, magnetometers, GPS, and light and proximity sensors, among others, are found in mobile devices, which manage gameplay, location reporting and the user experience.

Automated greenhouse systems use humidity, moisture, temperature, water, pH and CO_2 level detectors to manage the greenhouse environment, controlled by a PC or even manually through a mobile app. The advantage of captured and retained data is that it can be analysed and compared to the yield data for crops grown and adjustments can be made for future crops.

And let's not forget the concept of autonomous vehicles, where the importance of reliable sensor data is essential to safety.

DATA FILES

Software developers commonly not only need to be aware of data sources but also need to know how to create them, read them and update them. Fortunately, most modern programming languages have ample support for most of these data sources. The worked examples that follow are written in Python 3.X but they could be replicated in many popular languages with little difficulty.

As RAM contents are lost when power is removed, the use of non-volatile data sources (which do not lose their data when power is switched off), such as disk-based files, is a common programming concept. Unfortunately, these come in many different shapes and sizes.

Unstructured text files

Unlike a structured text file, these files typically contain content which cannot easily be converted or parsed into discrete fields of data.

Typical example of unstructured text files include:

- email bodies;
- SMS (short message service) texts;
- user feedback form text;
- audio transcriptions;
- lists;
- transaction logs.

Although the contents of the file are easy to read, without a set structure it is difficult to reliably extract key data. Typically, the software developer must create algorithms to read and parse the data successfully, knowing that any slight change to the content and/or layout may cause their code to fail.

Consider the following unstructured file, which represents an extracted sales transaction log:

```
1/10/20 Product A qty 30
01/10/20 Product B 20 A JONES
01/10/20 Product C 10 B GREEN
02/10/20 Product A qty 40
02/10/2020 Product A qty 40
```

As you can see, the format of the data appears unclear. There appears to be:

- a date (although this isn't consistently formatted);
- a product type (A, B or C);
- a quantity (sometimes prefixed by the abbreviation 'qty', sometimes not);
- an optional person's name (we don't know if this is the buyer, seller or manufacturer!).

As you can imagine, processing this type of data is awkward and often relies on leveraging additional technology, such as regular expressions (shortened as regex), to try to parse the unstructured and inconsistent data. Regular expressions are methods used in programming to match patterns such as strings or extracts of text.

The following Python 3.X extract demonstrates a possible solution.

```
import re
pattern = "^([0-9]{1,2}\/[0-9]{1,2}\/[0-9]{1,4}) Product ([ABC]) (?:qty )?([0-9]+)(?: |$)([A-Z ]*)$"

with open("odd_data.txt", "r") as my_data:
    for each_line in my_data:
        extracted_data = re.match(pattern, each_line, re.I)
        if extracted_data:
            print(extracted_data.groups())
```

As you can see, there's a lot of work involved, including a fairly convoluted regular expression, although this could be optimised.

The resulting data is now stored in a reasonably uniform data structure, ready for further processing.

```
('1/10/20', 'A', '30', '')
('01/10/20', 'B', '20', 'A JONES')
('01/10/20', 'C', '10', 'B GREEN')
('02/10/20', 'A', '40', '')
('02/10/2020', 'A', '40', '')
```

This type of data-processing nightmare can be improved by using either structured text files or a database.

Structured text files

Developers often need to communicate data between two different applications. The use of non-volatile structured text files can greatly assist this form of common data exchange. As mentioned, the popular ones to recognise are:

- **CSV:** the oldest and most commonly supported format.
- **XML:** often used to describe documents, configurations and web-based API requests. It can be difficult to process and can be susceptible to malicious XXE (XML external entity) injection attacks, which are listed in the OWASP (Open Web Application Security Project) top 10 web application security risks (OWASP 2017).
- **JSON:** anecdotally, the modern developer's favourite, typically used for web-based API responses as it closely mimics native object structures. It is easy to convert between objects and their JSON text-based representations (this process is called 'serialisation/deserialisation').
- **YAML:** often used for configuration files.

Many modern programming languages have either native support or third-party libraries which deal with the creation and parsing of these data formats. Always check these out before attempting to write your own solution!

Table 9.1 shows examples of popular structured text file formats.

Table 9.1 Structured text file formats

CSV	YAML

"name","age","gender","last_
attendance"
"Pick",19,"Female","01-11-2019"
"Jack",16,"Male","21-10-2019"
"Dom",19,"Female","03-11-2019"
"Jill",16,"Female","01-10-2019"
	--- 	
	-
	  name: Pick
	  age: 19
	  gender: Female
	  last_attendance: 01-11-2019
	-
	  name: Jack
	  age: 16
	  gender: Male
	  last_attendance: 21-10-2019
	-
	  name: Dom
	  age: 19
	  gender: Female
	  last_attendance: 03-11-2019
	-
	  name: Jill
	  age: 16
	  gender: Female
	  last_attendance: 01-10-2019

Table 9.1 (Continued)

JSON

```
[
 {
  "name": "Pick",
  "age": 19,
  "gender": "Female",
  "last_attendance": "01-11-2019"
 },
 {
  "name": "Jack",
  "age": 16,
  "gender": "Male",
  "last_attendance": "21-10-2019"
 },
 {
  "name": "Dom",
  "age": 19,
  "gender": "Female",
  "last_attendance": "03-11-2019"
 },
 {
  "name": "Jill",
  "age": 16,
  "gender": "Female",
  "last_attendance": "01-10-2019"
 }
]
```

XML

```
<?xml version="1.0"
encoding="UTF-8"?>
<people>
 <row>
  <name>Pick</name>
  <age>19</age>
  <gender>Female</gender>
  <last_attendance>01-11-2019</last_
attendance>
 </row>
 <row>
  <name>Jack</name>
  <age>16</age>
  <gender>Male</gender>
  <last_attendance>21-10-2019</last_
attendance>
 </row>
 <row>
  <name>Dom</name>
  <age>19</age>
  <gender>Female</gender>
  <last_attendance>03-11-2019</last_
attendance>
 </row>
 <row>
  <name>Jill</name>
  <age>16</age>
  <gender>Female</gender>
  <last_attendance>01-10-2019</last_
attendance>
 </row>
</people>
```

Each data format has its relative advantages and disadvantages. However, it is fair to say that JSON is probably the developer's favourite format (because, as noted, it mirrors the internal storage model of an object) while CSV (being the oldest) is the most ubiquitous (and therefore portable).

A software developer may not actually have a say in the data format being used. Instead, they may be presented with a file that has been exported from another application or service and have to work out how to access its data correctly – this is a fairly common software development task, especially in an enterprise pipeline. Table 9.2 shows examples of code used to read various file formats.

Table 9.2 Reading different file formats

<table>
<tr><th>Reading CSV</th><th>Common output</th></tr>
<tr><td>

```
import csv

with open('students.csv','r') as csv_file:
    csv_reader = csv.DictReader(csv_file)
    for row in csv_reader:
        for field in row:
            print(f'{field}: {row[field]}')
```

Reading YAML

```
import yaml

with open('students.yaml') as yaml_file:
    students = yaml.load(yaml_file, Loader=yaml.FullLoader)

    for student in students:
        for key, value in student.items():
            print(f'{key}: {value}')
```

Reading JSON

```
import json

with open('students.json') as json_file:
    students = json.load(json_file)

    for student in students:
        for key, value in student.items():
            print(f'{key}: {value}')
```

Reading XML

```
import xml.etree.ElementTree as ET

tree = ET.parse('students.xml')
students = tree.getroot()

for student in students:
    for field in student:
        print(f'{field.tag}: {field.text}')
```

</td><td>

```
name: Pick
age: 19
gender: Female
last_attendance: 01-11-2019
name: Jack
age: 16
gender: Male
last_attendance: 21-10-2019
name: Dom
age: 19
gender: Female
last_attendance: 03-11-2019
name: Jill
age: 16
gender: Female
last_attendance: 01-10-2019
```

</td></tr>
</table>

One of the key advantages of using Python as a programming language is its comprehensive support for structured data files 'out of the box'. The short extracts in Table 9.2 demonstrate standard library support for the four most popular types of structured data. Other programming languages generally aren't so well supported.

Binary files

When dealing with binary files, software developers are generally interested in the byte patterns contained in the file rather than their interpretations as Unicode (or ASCII) characters. Some applications output data files in a binary format. Alternatively, a developer might try to identify the electronic signature of a file to verify its true file type, or they might need to open a compressed archive.

The following code sample, written in Python for brevity, examines a specified folder of files and reports which appear to be in a .zip format. It does this by matching the various hexadecimal zip signatures of each file using the first four bytes of each. Successful matches are added to a file and then output once the scan is complete.

```
import binascii
import pprint
from os import listdir
from os.path import isfile, join

#specify a path to check its files
check_path = 'C:/Downloads/'

#get list of files, excluding sub-directories
list_files = [join(check_path, f) for f in listdir(check_path)
                                    if isfile(join(check_path, f))]

#store the hexadecimal signatures of first 4 bytes for zip files
zip_sig = (binascii.unhexlify(b'504B0304'), binascii.unhexlify(b'504B0506'),
           binascii.unhexlify(b'504B0708'))

#now, scan the files!
print(f"Starting to Scan {len(list_files)} files...")
zip_files = []

#for each file
for filename in list_files:
    print(f"Checking file {filename}...")

    #open each file and check first 4 bytes
    with open(filename, 'rb') as file_to_check:
        signature = file_to_check.read(4)
        #does it match the possible signatures?
        if signature in zip_sig:
            #yes, so append it to our list
            zip_files.append(filename)

#print the identified zip files
pp = pprint.PrettyPrinter(indent=4)
pp.pprint(zip_files)
```

Being able to read binary data from a backing storage device (any non-volatile data storage device, such as hard disk or USB) is a key skill for a software developer, particularly one interested in more low-level systems programming.

DATABASES

Connecting to databases is a bread-and-butter activity for most software developers and it is a particularly key activity for those working in Big Data, data science or commercial web applications.

In general, databases are either full-blown SQL-oriented relational database management systems (RDBMSs) or NoSQL alternatives. A business won't necessarily exclusively use one or the other – there's no reason why both can't be used in some capacity, depending on the nature of the data. However, their designs, usage and methods of interaction are very different.

Let's look at each in turn from a developer's perspective.

Mainstay 'heavyweight' relational database management systems

In terms of RDBMSs, most developers are likely to encounter one of the following:

- **MySQL:** an industry mainstay, now owned by Oracle and used by the likes of Facebook, GitHub, PayPal, Twitter and YouTube.
- **MariaDB:** a community-developed, open source fork of MySQL, released under the GNU General Public License and used by the likes of DBS Bank, Google, Mozilla and Red Hat.
- **PostgreSQL:** a mature, free and open source database supported by multiple operating systems.
- **Microsoft SQL Server:** a mature RDBMS commercial product created by Microsoft and available in many differently scaled editions depending on business need (including Azure – cloud-based – support).
- **Oracle:** a RDBMS created using scalable architecture produced by the Oracle Corporation and particularly used by larger organisations which process data across the world using Oracle SQL.

All of these RDBMS use SQL for managing, manipulating and accessing relational data contained in databases and tables. As such, they are designed to store structured data with complex relationships between different entities, for example:

- **One-to-one relationships:** a person has one customer account.
- **One-to-many relationships:** a customer account may have many orders.
- **Many-to-many relationships:** an author can write many books and a book can be written by many authors. This type of relationship usually makes use of a resolving/link/junction entity.
- **Self-referencing relationships (also known as recursive):** rows in a table are related to other rows in the same table.

SQL-based RDBMSs traditionally require a predefined schema to describe their logical and physical data structure. This schema details the databases and tables as well as their relationships, keys, indexes, stored procedures, user permissions and so on. The resulting complex data structures support the most intricate data interrogation queries imaginable and there are many published texts available to support your learning of them.

These types of RDBMS typically run as resource-hungry background services (or 'daemons'), allowing simultaneous connections from multiple clients and capable of processing complex queries at incredible speed. For this reason, they can typically be found at the very heart of any modern commercial web application, and indeed MariaDB and MySQL typically represent the 'M' in any LAMP (Linux, Apache, MySQL and PHP), WAMP (Windows, Apache, MySQL and PHP) or MAMP (Mac, Apache, MySQL and PHP) stack of technologies.

Modern solutions often involve cloud-based implementations of virtualised RDBMS servers. One of the key requirements of this type of RDBMS is its support for ACID (atomicity, consistency, isolation and durability) – a set of properties which ensure that any database transaction is either completed successfully or 'rolled back' to its previous state in the event of errors. This guarantees the data's integrity.

In simple terms:

- **Atomicity:** each transaction is treated as an atomic unit; **all operations are completed, or none are**.
- **Consistency:** the result of a transaction **must** leave the database in a consistent state once completed.
- **Isolation:** all transactions are carried out **separately**, simultaneously and in parallel, affecting no other.
- **Durability:** all outstanding transactional updates are completed even if the service is unexpectedly disrupted or restarted – they **must** be finished.

SQLite

Unlike the mainstay RDBMSs listed, SQLite is a RDBMS contained with a C library and does use the client–server model. It is embedded into the application itself. Although it is ACID compliant, it can have issues with data type conversions. However, it remains a popular choice for embedded databases for use in applications. As it is built into many web browsers for maintaining local storage, it is fair to say that it is probably one of the most frequently used RDBMS, even if its embedded existence is not always obvious.

It also has 'bindings' ('wrapper' libraries which allow different technologies to be used in concert) for many popular programming languages, including C, C#, C++, Java, JavaScript, PHP and Python, making it a popular choice for developers. Additionally, it is built into popular web application frameworks such as Django, Laravel and Ruby on Rails. Several CMSs, such as Drupal and WordPress, also make extensive use of it.

Working with a RDBMS

The following example demonstrates a short section of Python which accesses data from an SQLite database called 'project' and a table called 'engineers'. You will note that the example demonstrates three parts of the SQL language:

- **SQL DDL (Data Definition Language):** when creating the table;
- **SQL DML (Data Manipulation Language):** when inserting new rows of data into the table;
- **SQL DQL (Data Query Language):** when performing the 'select' statement.

The resulting record set (a collection of rows that match the query) is then iterated using a **for** loop and printed on screen.

```
import sqlite3

# create a connection to the database
conn = sqlite3.connect('project.db')

# create a cursor
c = conn.cursor()

# create engineer table
c.execute('CREATE TABLE engineers (name text, location text, salary real)')

# insert a row of data
c.execute('''INSERT INTO engineers VALUES
                     ('Wein','Dept A', 50000.0),
                     ('Mantlo','Dept B', 60000.0),
                     ('Dillon','Dept A', 70000.0)''')

# save the changes by performing a commit
conn.commit()

# create a query parameter
param = (50000.0,)

# execute the query
results = c.execute('SELECT * FROM engineers WHERE salary>?', param)
for row in results:
    print(row)

# close the database connection (good practice)
conn.close()
```

The resulting output from this short Python script is:

```
('Mantlo', 'Dept B', 60000.0)
('Dillon', 'Dept A', 70000.0)
```

Security concerns

It would be remiss not to raise the spectre of security concerns connected to applications which process SQL queries. The most obvious and often criminally overlooked task is to sanitise potentially harmful input.

This vulnerability is colloquially referred to as 'SQL injection' and it is recommended that you visit the OWASP website (https://owasp.org) to read further on this topic, including various mitigation techniques. SQL injection is an aggressive hacking technique which exploits vulnerabilities in programming code by allowing arbitrary SQL code to be run by modifying existing SQL statements. The most common attack vector occurs when HTML form-based inputs are not correctly sanitised before being processed.

In the SQLite example provided above, 'parameter substitution' (note the '?' used as a placeholder) is used rather than manually concatenating the string. This technique is generally much safer because parameter values are often 'cleaned' by the underlying API's functions.

Frankly, software developers have very little excuse if their applications are found to be subject to SQL injection attacks as many modern development frameworks and most programming language libraries have ample tools to prevent such issues (e.g. cleaner functions and parameterised queries). Despite that, as of 2017 (the most recent OWASP update at the time of writing; see OWASP 2017), SQL injection remains the number one vulnerability and its contribution to many serious data breaches (known and unknown) worldwide cannot be underestimated.

'Lightweight' databases (e.g. NoSQL)

Contrary to some sources, 'NoSQL' does not mean 'there's no SQL' – it means 'not only SQL'.

Instead of using the normal 'heavyweight' database schema, NoSQL uses these four 'lighter' data structures:

- **key-value:** uses an associative array (or dictionary) consisting of key-value pairs;
- **wide column store:** a two-dimensional key-value store where keys and value formats can vary from row to row in the same table;
- **graph:** as the name suggests, data is represented using graphs (the mathematic structure that models relations between objects);
- **document:** data is encoded in structured text formats including JSON, XML and YAML.

Common NoSQL databases include:

- Redis (key-value-based);
- MongoDB (document);
- Elasticsearch (document);
- Microsoft Cosmos DB (document);
- Amazon DynamoDB (wide column store);
- Neo4j (graph).

Many programming languages support connections to such NoSQL databases using appropriate APIs. For example, Python developers can choose MongoDB as a database solution or use Amazon's DynamoDB in the Amazon Web Services cloud.

Working with NoSQL

The following example demonstrates a short section of Python 3.X which accesses data from a NoSQL MongoDB database called 'project' and a table called 'engineers'.

```
import pymongo

#connect to the MongoDB server
myclient = pymongo.MongoClient("mongodb://localhost:27017/")

#use the database or create if doesn't exist
mydb = myclient["project"]

#use the collection (table) or create if doesn't exist
mycol = mydb["engineers"]

#insert new documents (rows)
mylist = [
           {"name": "Wein", "location": "Dept A", "salary": 50000.0},
           {"name": "Mantlo", "location": "Dept B", "salary": 60000.0},
           {"name": "Dillon", "location": "Dept A", "salary": 70000.0}
]
ins_objs = mycol.insert_many(mylist)

#query all documents (rows) in a collection (table) using an expression
results = mycol.find({"salary":{"$gt":50000.0}})
for row in results:
    print(row)
```

Note the absence of a traditional **SQL SELECT** statement to query the data, or indeed an **INSERT** statement to add new data.

The resulting output from this short Python script is:

```
{'_id': ObjectId('5e170a0cf4058a1e80d680a5'), 'name': 'Mantlo',
'location': 'Dept B', 'salary': 60000.0}
{'_id': ObjectId('5e170a0cf4058a1e80d680a6'), 'name': 'Dillon',
'location': 'Dept A', 'salary': 70000.0}
```

In this example, the returned data is stored in dictionary objects, including an auto-generated **_id** field which makes each document (or row) unique.

Internally, MongoDB stores this collection (table) in JSON. It can be viewed using appropriate development tools, such as MongoDB Compass:

```
{
    "_id": {
        "$oid": "5e170a0cf4058a1e80d680a4"
    },
    "name": "Wein",
    "location": "Dept A",
    "salary": {
        "$numberDouble": "50000"
    }
}
```

```
{
    "_id": {
        "$oid": "5e170a0cf4058a1e80d680a5"
    },
    "name": "Mantlo",
    "location": "Dept B",
    "salary": {
        "$numberDouble": "60000"
    }
}

{
    "_id": {
        "$oid": "5e170a0cf4058a1e80d680a6"
    },
    "name": "Dillon",
    "location": "Dept A",
    "salary": {
        "$numberDouble": "70000"
    }
}
```

NoSQL: alternatives to SQL statements

You may have noticed that the sample code does not use a traditional SQL statement to query the data. Instead a **`find`** method is used:

```
results = mycol.find({"salary":{"$gt":50000.0}})
```

In this case, we need to specify the field (**`salary`**), the relational operator (**`$gt`**, which means 'greater than' or '>') and the value (**`50000.0`**).

Variations of this type of 'no SQL SELECT' function can be found in many programming language frameworks, offering more protection against SQL injection attacks.

As before, only documents that meet this criterion are returned and these are iterated via the use of a **`for`** loop and printed on screen.

Which to use: SQL or NoSQL?

Essentially, the choice of whether to use SQL or NoSQL comes down to scale and complexity of data. If the data is simple and mostly non-relational, NoSQL is acceptable. Larger data sets with complex relationships really require a traditional SQL-based RDBMS server. Conversely, the type of raw data used in big data sets ideally suits NoSQL.

WEB-BASED APPLICATION PROGRAMMING INTERFACES

Organisations often rely on external data sources in order to solve a software development problem. A common example is UK postcodes and looking up their associated addresses. Once the sole province of the Royal Mail, there are now many online services which offer access to this type of data through a web-based API (application programming interface). Well-documented APIs allow a developer to quite painlessly access data over the internet. This data can then be used in their own applications, for example to auto-complete a shipping address (which minimises the need for time-consuming user input).

Accessing this type of service is a pretty common task for a developer. Let's see how this works using an example.

Figure 9.1 Postcodes.io's web-based API (© Ideal Postcodes, https://github.com/ideal-postcodes/postcodes.io/blob/master/LICENSE)

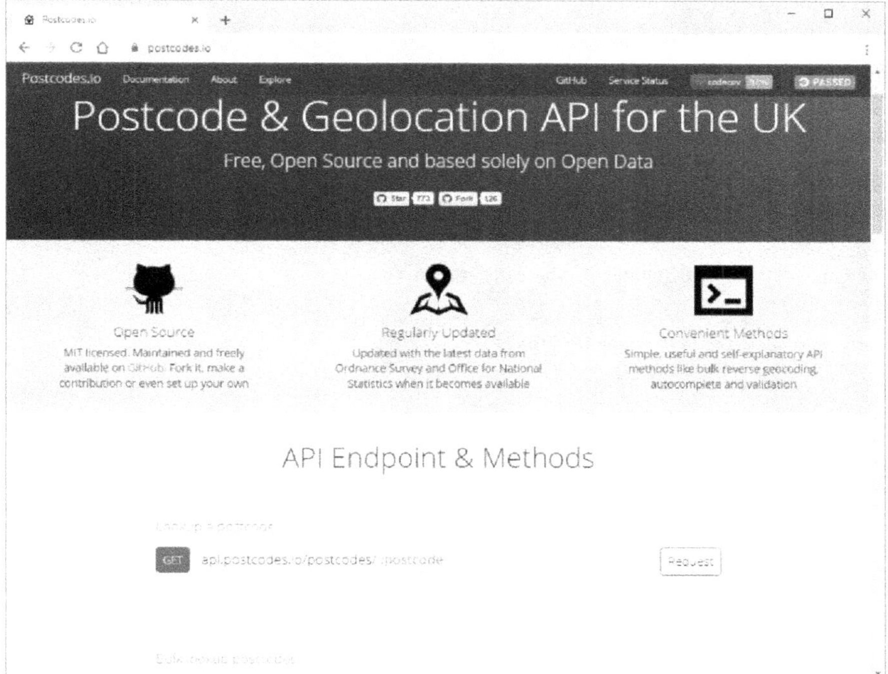

The website Postcodes.io offers a free (and open source) postcode and geolocation API for the UK, as shown in Figure 9.1. In order to access the API as a developer, we need to know:

- the available endpoints – the URLs and variables required;
- the HTTP methods/verbs (i.e. **`GET`**, **`POST`** etc.);
- the data format (and contents) of the API's response;
- whether the protocol being used is HTTP or HTTPS (i.e. secure).

The most basic interaction for this API is the following **`GET`** request, sent using HTTP:

api.postcodes.io/postcodes/<postcode>

This will result in a JSON response consisting of:

- a HTTP status code (e.g. **200** for 'OK');
- a result set containing geolocation information about the given postcode.

Creating a short code extract to access this type of data is relatively straightforward. The following example was coded in Python 3.X (mainly for brevity) but could have been implemented in most commercially available programming languages:

```
import requests
from pprint import pprint

#create the API request endpoint
endpoint = "http://api.postcodes.io/postcodes/"
postcode = "SN21FA"

endpoint += postcode

#make the request
new_request = requests.get(endpoint)

#handle the response
if new_request.status_code == requests.codes.ok:

    #decode and print the response (we could do more)
    my_response  = new_request.json()
    pprint(my_response)
else:
    print("No response from API")
```

The following JSON response is received from the web API request:


```
{'result': {'admin_county': None,
       'admin_district': 'Swindon',
       'admin_ward': 'Rodbourne Cheney',
       'ccg': 'NHS Swindon',
       'ced': None,
       'codes': {'admin_county': 'E99999999',
            'admin_district': 'E06000030',
            'admin_ward': 'E05008967',
            'ccg': 'E38000181',
            'ccg_id': '12D',
            'ced': 'E99999999',
            'nuts': 'UKK14',
            'parish': 'E04012663',
            'parliamentary_constituency': 'E14000851'},
       'country': 'England',
       'eastings': 414690,
       'european_electoral_region': 'South West',
       'incode': '1FA',
       'latitude': 51.565685,
       'longitude': -1.789463,
```

```
'lsoa': 'Swindon 012B',
'msoa': 'Swindon 012',
'nhs_ha': 'South West',
'northings': 185222,
'nuts': 'Swindon',
'outcode': 'SN2',
'parish': 'Central Swindon North',
'parliamentary_constituency': 'North Swindon',
'postcode': 'SN2 1FA',
'primary_care_trust': 'Swindon',
'quality': 1,
'region': 'South West'},
'status': 200}
```

The JSON response is well structured and therefore easily processed in Python as a dictionary data structure (like an associative array in other languages).

It is worth noting that the Requests module being imported at the start of the Python script had to be installed via 'pip' (its default package management system) as it is not part of Python's standard library. It is the ability for developers to use sophisticated pre-written, reliable and well-tested libraries like this that takes much of the heavy lifting out of the everyday tasks they encounter.

Most modern programming languages support the generation of HTTP1.1 requests, and, although the syntax will undoubtedly differ, the same steps should be followed:

1. build the request;
2. make ('fire') the request;
3. check the status of the response;
4. decode the response.

Modern web APIs often rely on a RESTful architectural style, a technique which you should be familiar with.

RESTful services

RESTful (representational state transfer) services are used to create web services by providing interoperability functionality between different systems on the internet (see the example in Figure 9.2).

RESTful operations tend to mirror the basic CRUD-style operations (i.e. create, read, update, delete) that a programmer might wish to perform on a database. RESTful operations are specified using a combination of URL and (optionally) POSTed data (see Table 9.3).

Figure 9.2 RESTful services

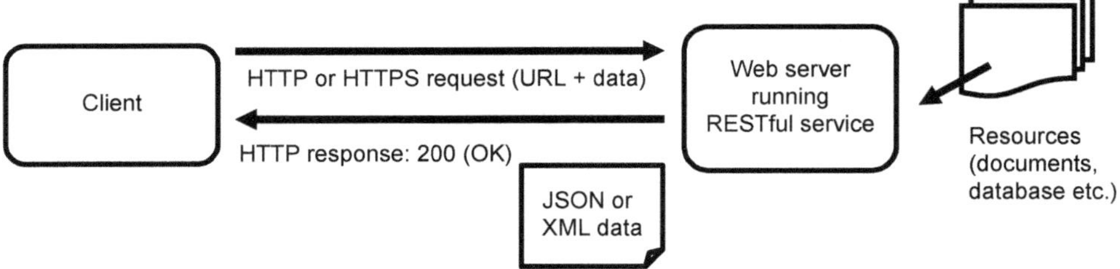

Table 9.3 RESTful interactions, demonstrating the uniformity of the request interface

Sample operation	HTTP method	URL	POST data (JSON)

Create customer	POST	**/myservice.com/**<customer_id>	[
			{
			"firstname": "john",
			"lastname": "smith",
			"dept": "accounting"
			}
			]

Read customer	GET	**/myservice.com/**<customer_id>	No posted data
Read customer**s**	GET	**/myservice.com/**	No posted data

Update customer	PUT	**/myservice.com/**<customer_id>	[
			{
			"dept": "development"
			}
			]

Delete customer	DELETE	**/myservice.com/**<customer_id>	No posted data

RESTful services are quick to 'spin up' (launch or instantiate) for a seasoned developer. A simple micro-framework such as the Python-based Flask can be up and running in an hour or two, particularly when paired with a NoSQL document-oriented database such as MongoDB (which essentially uses JSON).

Ultimately these types of commercial web solution tend to use more featured MVC (model-view-controller) frameworks (such as Laravel), workhorse server-side programming languages (such as PHP or ASP.NET) and full RDBMSs (such as MS SQL, MySQL, MariaDB or PostgreSQL).

TIPS WHEN WORKING WITH DATA

Being aware of potential pitfalls when working with data is essential. Here are some examples of things you should consider:

- **Never** trust data, even if from a verified source – always validate and sanitise it **before** it is either processed or stored.
- Be aware of the different data formats, particularly structured file formats such as CSV, JSON and XML.
- A working knowledge of regular expressions can be very useful for a software developer.
- Be careful when constructing SQL interactions with a database; ensure that malicious data (or code) cannot be injected.
- When testing coded solutions or databases, always use realistic data.
- Back up data regularly.
- Carefully observe the principles of the GDPR when dealing with data.
- Always think ahead when choosing how to store data. Using constrained formats early in development may create issues in the future when software needs to be maintained (e.g. making a customer number numerical initially and then wanting to add alphabetical characters later on).

SUMMARY

When software developers learn to program, they typically build simple, 'message pair'-style applications which execute at the console, for example:

```
(OUTPUT) What is your name?
(USER INPUT) Tom
(OUTPUT) Hello, Tom!
```

However, not all user input is received from the keyboard; developers must be familiar with techniques that accept input from a variety of sources (files, networks, databases, devices etc.). This harkens back to the development of Unix and its core design notion that each 'program' should have one job, do it well and not expect input from any singular source. In being able to deal with different sources of data, the developer follows that early guiding principle. In this chapter we have given you a broad overview of the possibilities available.

The next chapter focuses on testing and the analysis of test results.

10 TESTING CODE AND ANALYSING RESULTS

'"How to test?" is a question that cannot be answered in general. "When to test?" however, does have a general answer: as early and as often as possible.'

– Bjarne Stroustrup

Making a better product in the future will depend on the developer knowing what is wrong with the existing one and learning lessons from the experience. The testing process is key to identifying issues with a product: this chapter provides insight into what is important about testing and test analysis.

OVERVIEW OF TESTING

Many developers will admit to seeing testing as the most tedious part of their job – after all, it's generally not as much fun as coding the solution. Nevertheless, it is almost certainly the most important part of the developmental cycle, principally because its goal is to ensure that the client's requirements have been fully met and that the resulting solution is:

- **Correct:** there are no defects; it is a quality product.
- **Robust:** the product copes with errors during execution and erroneous (potentially malicious) user input.
- **Reliable:** the results are consistent.
- **Efficient:** the product responds in a timely manner and uses resources efficiently.

Before we investigate testing itself, let's review the primary difference between quality assurance and testing as these two terms are often used interchangeably.

The difference between testing and quality assurance

Often confused, both testing and quality assurance are components of development that aim to deliver a stable product that meets the needs of users and is free of errors.

'Tests are sometimes mistaken with quality assurance. These two notions are not identical: 1) quality assurance ensures that the organization's processes are implemented and applied correctly; 2) testing identifies defects and failures, and provides information on the software and the risks associated with their release to the market.'

—Bernard Homès

Table 10.1 considers the activities involved in both testing and quality assurance to demonstrate the differences between the two concepts.

Table 10.1 Comparison of testing and quality assurance activities

Testing	Quality assurance
Explores the code to uncover **bugs** and **defects** which need to be resolved.	Considers the product and how well it **conforms** to the original **requirements**. Where appropriate, this may also include checking conformity with formal quality standards, such as ISO 9000 or the Capability Maturity Model (CMM). ISO 9000 is an industry-agnostic set of standards intended to ensure the maintenance of an effective quality-assurance system. The CMM is a framework for organising continuous process improvement.
Examines the **product** or **outcome** of the activity to check that it **works**.	Examines the **processes** undertaken to create the product or outcome.
Aims to make the **client** and **users happy**.	Aims to **improve future developments**.
Generates activity which is subsequently **reactive**.	Generates activity which is subsequently **proactive**.
Undertaken by **testers**.	Undertaken by the whole **project team.**

Passing all the functional and non-functional tests is not a guarantee of a quality product.

Let's explore some of the basic approaches to successful testing.

Testing as part of the V-model methodology

The V-model is a different type of model from the ones mentioned in Chapter 4 and it represents the development of a software solution. It bears some similarities to the Waterfall model in that there is a sequential path of processes – one stage must be completed satisfactorily before the next may commence. However, where the difference is noticeable is in the V-model's treatment of testing.

In the Waterfall model, testing isn't commenced until coding has been completed. This is quite late in the development cycle.

In the V-model (as shown in Figure 10.1), testing of the product is planned in parallel with each of its corresponding development phases. In other words, there are actually two cycles – one for the developer (called the verification phase) and one for the tester (called the validation phase). It is these dual phases, verification and validation, that give this development model its distinctive name.

Figure 10.1 Software development using the V-model

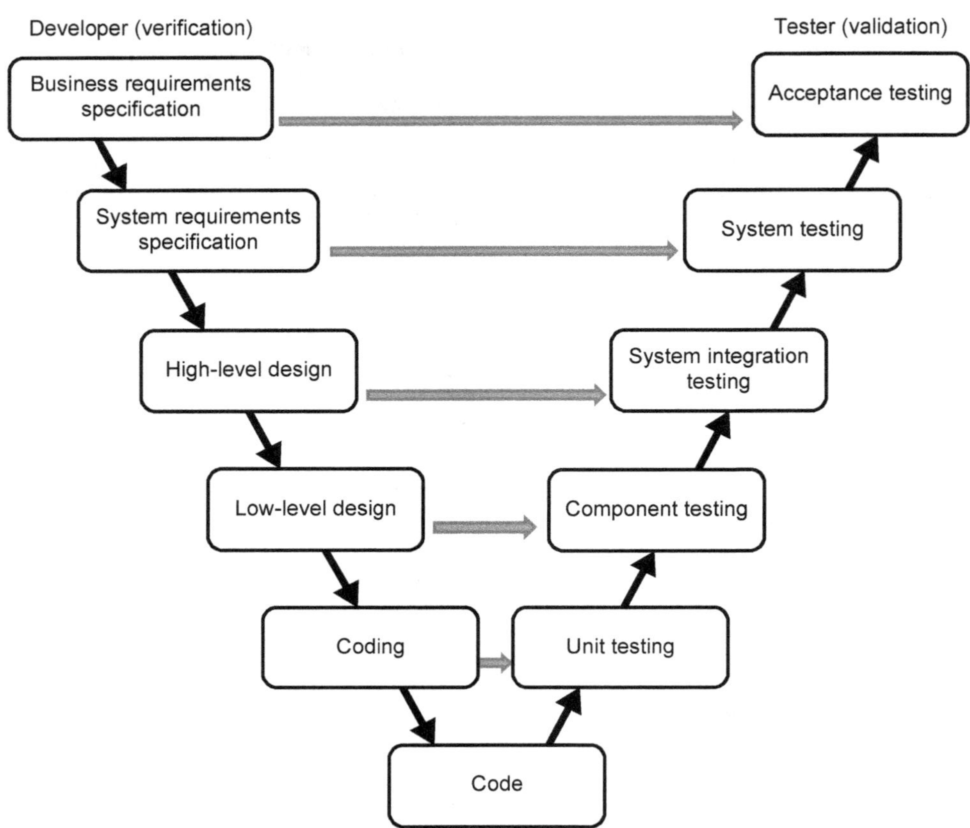

As you can see from Figure 10.1, each developer stage has a corresponding tester stage which it informs – for example, the low-level design involves designing the classes, functions and so on which form the components of the target solution, which will be physically coded in the next sequential phase. However, at the same time, this low-level design will inform the component tests that are eventually used in the validation flow.

The core advantage here is the early development of test products, as they will emerge from the developer sequence in a natural and timely way. It's not an ideal approach, of course – there's no scope for early prototyping as you would see in a RAD (rapid application development) approach, and there is no MVP (minimum viable product) as in an Agile approach. For this reason, the V-model is seen as a somewhat rigid lifecycle model and one that is best suited to small or medium projects where the scope and complexity of the task are well understood.

METHODS OF TESTING

Although code extracts will have been tested throughout development, formal testing is undertaken to check that the original development requirements have been met. The aim of testing is also to prove that the solution is correct by attempting to break it in some way. Passing tests is the developer's way of convincing the user and client that the software is not incorrect.

Because the testing phase **traditionally** (though not always) occurs late in the software development lifecycle, some development teams may be tempted to leave the testing activities until after the coding has been completed. However, modern practice is to carry out continuous testing throughout the coding activity in the form of unit tests. This will ensure that errors will be found and can be resolved quickly.

The essence of testing is to identify bugs in the software that occur when calculations are made, data files are parsed or users interact with the software – in theory, anywhere in the executable lifetime of an application.

In truth, any interaction with an external data source represents low-hanging fruit for testing, principally because it introduces data that cannot (and should not) be explicitly trusted; for example, a failure to validate user inputs effectively could produce unexpected results during use. Other common bugs include calculation errors based on poor application of logic, potential division-by-zero errors, the use of incorrect mathematical formulae, and uncaught mismatches in data types. Anything is possible.

Testing can never hope to catch everything or prove beyond a shadow of a doubt that an application is 100% fault free – but it must to do its best to reassure us.

White box testing

White box testing is a category of testing where the developed code is known to the testing team. Its role is to check that the flow of user inputs, processes and outputs operates correctly when the application is run.

The advantages of white box testing, put simply, are:

- It can be started very early in the development phase – as soon as the first functions are written.
- It can be automated.
- It often highlights opportunities for code refactoring (see Chapter 7), making efficiencies and tidying the codebase.
- It is normally very thorough as it is possible to ensure that all lines of executable code are tested.

Often this type of testing is performed by the developers – although this has advantages and disadvantages. On the plus side, the developers fully understand their code, can fix errors quickly and have a vested interest in ensuring that the code is working correctly. Conversely, developers can suffer from confirmation bias – a subconscious cognitive condition where tests may be favoured that prove the software works as the developers expect (see more on confirmation bias later in this chapter). For this reason, developers outside the core software development team may be used. Practices can vary across the IT sector.

There are many different testing techniques that can be used to achieve white box testing. Let's examine a few.

Code coverage

Code coverage is a popular and traditional metric used to describe the amount of code that is executed when a selected set of tests is run.

It is usually expressed as a percentage, often drilled down to each module or function. Ideally, the coverage should be as close to 100% as is practically possible. Coverage should minimally include:

- **functions** and their **parameters**;
- individual **statements**;
- **all branches**, for example **`IF...ELSE`** statements which offer multiple logic pathways (decision-to-decision paths), and correct execution of loop constructs based on a series of test conditions.

Let's create a relatively simple example that provides suitable scope to demonstrate the concept of code coverage. We'll use prime numbers (numbers greater than 1 that cannot be evenly divided by any number other than themselves or 1).

The following Python 3.X module (primeNumber.py) has two functions:

- The first function, **`is_prime()`**, tests whether a given number is a prime number, returning a Boolean (true/false) value.
- The second function, **`next_prime_available()`**, generates the next prime number available after a given number.

```
def is_prime_number(value: int) -> bool:
    """Check to see if given value is prime."""
    if value > 1:
        for number in range(2, value):
            result = value % number
            if result == 0:
                return False
        return True
    return False


def next_prime_available(value: int) -> int:
    """Get next prime available, larger than value."""
    index = value
    while True:
        index += 1
        if is_prime_number(index):
            return index
```

The next Python code extract (primeRun.py) is used to test numbers between 1 and 20 (exclusive) by using the **is_prime()** function. This is known as a 'testrunner'.

```
#tester for coverage module
from primeNumber import is_prime_number

for counter in range(1, 20):
    if is_prime_number(counter):
        print(f"{counter} is prime")
    else:
        print(f"{counter} is not prime")
```

When the test code is run, Python's coverage unit keeps track of each line of code that has been executed – and those that have not. If the code makes use of library functions, these too may be included in the coverage statistics. Here is the sample process executed line by line at the Python shell:

```
>>> import trace

>>> my_trace = trace.Trace(trace=1, count=1, timing=True)

>>> my_trace.run('import primeRun')

>>> my_results = my_trace.results()

>>> my_ressults.write_results(show_missing=True, summary=True,
coverdir=".")
```

The Python process creates summary statistics per module:

```
lines cov% module (path)

 14  50% primeNumber (C:\Users\User1\AppData\Local\Programs\Python\
Python37-32\primeNumber.py)

  5 100% primeRun (C:\Users\ User1\AppData\Local\Programs\Python\
Python37-32\primeRun.py)

 439  14% rpc (C:\Users\User1\AppData\Local\Programs\Python\
Python37-32\lib\idlelib\rpc.py)

 364  0% run (C:\Users\User1\AppData\Local\Programs\Python\Python37-32\
lib\idlelib\run.py)
 597  4% threading (C:\Users\User1\AppData\Local\Programs\Python\
Python37-32\lib\threading.py)

 450  0% trace (C:\Users\User1\AppData\Local\Programs\Python\
Python37-32\lib\trace.py)
```

It also creates inspectable 'coverage files' which have a .cover extension. Here is the coverage file for the **primeNumber** module which contains the functions:

```
       def is_prime_number(value):
           """Check to see if given value is prime."""
   19:     if value > 1:
   81:         for number in range(2, value):
   73:             result = value % number
   73:             if result == 0:
   10:                 return False
    8:         return True
    1:     return False

>>>>>> def get_next_prime(value):
           """Get next prime available, larger than value."""
>>>>>>     index = value
>>>>>>     while True:
>>>>>>         index += 1
>>>>>>         if is_prime_number(index):
>>>>>>             return index
```

As you can see, the numbers alongside each line of Python code indicate the number of times each has been executed by the test script. It is good to see that every **return** statement, each representing a separate logic pathway, has been tested successfully.

In addition, any line of code not executed by the test script is flagged using the **>** (show missing) chevrons. Of course, the test script does not make use of the second function at all! Consequently, the **primeNumber** module reports 50% coverage; our test script must be improved! How? Simply by making sure that the testrunner makes use of **both functions**, not just the first one.

Similar open source coverage tools exist for many popular programming languages, including C, C++, Java and PHP. Commercial applications such as Microsoft Visual Studio Enterprise have such facilities fully integrated.

Unit testing

Unit testing is a common software testing technique which you are likely to encounter in the industry. Here you are testing the smallest components of a program using white box principles; in a procedural solution this may be a function or procedure, whereas in object-oriented programming it is likely to be a class method.

The role of a unit test is to check the accuracy and performance of an individual component, with the key question being: 'Does it work as expected?' Such tests are usually written by the programmer or a member of the software development team but may sometimes be written by independent software testers. An internal arrangement is usually preferred, though.

So why use unit tests?

- They make it quicker to locate and fix (debug) bugs within smaller components in a larger codebase.
- Quicker debugging is obviously less expensive.
- They avoid placing debug code (a familiar but inadvisable programming habit) into a live codebase (something which can be exceptionally dangerous) by effectively isolating the testing (and the test code).
- They are reusable resources and can easily be re-run when a component is modified in the future, thereby immediately highlighting the introduction of any 'broken' code whenever an outcome is different (i.e. 'It was working OK before but now it isn't – what was changed?').

Let's consider a simple but practical example. The following Python 3 function tests whether an integer is a prime number:

```
def is_prime_number(value: int) -> bool:
    """Check to see if given value is prime."""
    if value > 1:
        for number in range(value):
            result = value % number
            if result == 0:
                return False
        return True
    return False
```

The most basic tests we could consider involve asking:

- Does the function identify prime numbers correctly?
- Does the function return the correct data type (a Boolean)?

In Python, a unit test is written as a separate class-based module which contains several methods which make separate calls to our function. Each method tests a different expectation by making an 'assertion' (a statement which we believe to be true).

For example, we believe that the function will always return a Boolean value (either true or false); that's an assertion. Our simple Python 3 unit test might look like this:

```
import unittest

# unit test should be testing one function
from primeNumber import is_prime_number


class PrimeUnitTest(unittest.TestCase):
    """Tests for primes.py module"""

    def test_return_type(self):
        """should return a Boolean"""
        self.assertIsInstance(is_prime_number(10), bool)
```

```
    def test_is_eleven_prime(self):
        """is eleven (11) revealed to be a prime?"""
        self.assertTrue(is_prime_number(11))

    def test_is_ten_prime(self):
        """is ten (10) revealed to be prime?"""
        self.assertFalse(is_prime_number(10))

    def test_special_one_is_prime(self):
        """is special case (1) revealed to be a prime? It should
        not be as it only has one positive divisor"""
        self.assertFalse(is_prime_number(1))

if __name__ == '__main__':
    unittest.main(verbosity=2)
```

This unit test makes four separate assertions which we will test by calling the function and checking the outcome:

- The function will always return a Boolean value.
- The function will return **True** if 11 is passed into the function.
- The function will return **False** if 10 is passed into the function.
- The function will return **False** if 1 is passed into the function (it's our special case and is treated separately).

Running the unit test produces the following output:

```
test_is_eleven_prime (__main__.PrimeUnitTest)
is eleven (11) revealed to be a prime? ... ok
test_is_ten_prime (__main__.PrimeUnitTest)
is ten (10) revealed to be prime? ... ok
test_return_type (__main__.PrimeUnitTest)
should return a Boolean ... ok
test_special_one_is_prime (__main__.PrimeUnitTest)
is special case (1) revealed to be a prime? It should ... ok
----------------------------------------------------------------------
Ran 4 tests in 0.170s
```

In this example, the unit test has called the function four times using the test values supplied. In each case, the return value (*hopefully* a Boolean **True** or **False**) is compared against the assertions made.

For example, this discrete test:

```
def test_return_type(self):
    """should return a Boolean"""
    self.assertIsInstance(is_prime_number(10), bool)
```

is making an asserting that a Boolean will be returned.

This test passes:

```
test_return_type (__main__.PrimeUnitTest)
should return a Boolean ... ok
```

Let's now modify the code and deliberately break it (the change is boxed):

```
def is_prime_number(value: int) -> bool:
    """Check to see if given value is prime."""
    if value > 1:
        for number in range(value):
            result = value % number
            if result == 0:
                return False
        return True
    return True
```

```
test_is_eleven_prime (__main__.PrimeUnitTest)
is eleven (11) revealed to be a prime? ... ok
test_is_ten_prime (__main__.PrimeUnitTest)
is ten (10) revealed to be prime? ... ok
test_return_type (__main__.PrimeUnitTest)
should return a Boolean ... ok
test_special_one_is_prime (__main__.PrimeUnitTest)
is special case (1) revealed to be a prime? It should ... FAIL
======================================================================
FAIL: test_special_one_is_prime (__main__.PrimeUnitTest)
is special case (1) revealed to be a prime? It should
----------------------------------------------------------------------
Traceback (most recent call last):
  File "test_primes.py", line 24, in test_special_one_is_prime
    self.assertFalse(is_prime_number(1))
AssertionError: True is not false
----------------------------------------------------------------------
Ran 4 tests in 0.145s
FAILED (failures=1)
```

Here the output results are achieved by re-running the same test on the now modified code:

The accidental modification has created an outcome where 1 is considered to be a prime number (**`True`** is returned). Our original, and importantly unchanged, unit test asserts that 1 should return **`False`**. That test now fails, indicating that the once-correct behaviour of the function has been broken by a recent modification.

Our test case should lead us back to the incorrectly modified line of code; the error is then quickly debugged and the test re-run (successfully).

This may appear to be a trivial example, but it quickly illustrates two basic points:

- Modification of existing code typically introduces new bugs.
- Testing such small components in a software solution permits targeted and efficient debugging.

Unit testing techniques vary from language to language; however, they generally follow the same principles. For instance, Python's technique would be very familiar to Java programmers, being quite similar to the popular JUnit testing framework.

For these reasons you should consider unit tests a core component of your role as a software developer.

Integration testing

Building on unit testing, the other main white box testing technique is integration testing, where groups of units are tested together.

Although units, modules, functions and so on may be tested independently, they will be combined with others to form a complete software solution. As such, it is important to test whether using them in concert raises any issues. This is where integration testing becomes important, as it uncovers issues that arise when the units interact with each other. Typical issues include:

- **Naming conflicts:** for example, two units or modules have functions with the same name, causing pollution of the application's namespace.
- **Flow issues:** for example, the output from one unit is not in the correct format to act as the required input of another. This is particularly problematic in the functional programming paradigm; individually they are fine, but as a processing pipeline they fall over.

Other types of white box testing

There are other types of testing available which rely on knowledge of the software development project's codebase:

- **Regression testing:** the act of re-running your tests to ensure that their behaviours have not altered after code modifications – for example, tests which failed still fail and successful tests are still OK.
- **Penetration testing ('pen testing'):** discovering weaknesses in implementation which can be exploited with a little technical know-how. For instance, poor user input validation may lead to system exploitation (buffer overflows or SQL injection attacks are common examples).
- **Memory leak testing:** where application performance and/or system stability is called into question, the possibility is often raised of memory leaks within the software solution. It is more prevalent in applications where memory is allocated dynamically by the developer and then not de-allocated once used (e.g. C, C++ solutions in particular) than in those which have better 'garbage collection' (i.e. those where memory is de-allocated after use, such as Java and Python) – although it can still happen.

Black box testing

Black box testing is a feature of ongoing development, as each component is tested and any issues corrected.

More formal black box testing is undertaken later in the development process. This is because with black box testing, the testers are not interested in the internal workings of the code (or the system) and will not be given access to them. The testing activity is generally used to check for the presence of expected functionality as in the original specification. All calculations, however, are extensively checked.

In addition, the activity establishes how the software will integrate with the other system components (system testing), how it performs when managing the expected volumes of data and how it copes with the anticipated number of users.

Further types of testing include the following:

- **Internal** and **external acceptance testing** are based on black box principles because the techniques seek to check that the software meets a range of expectations without any emphasis on the construction of the code.
- **Internal acceptance testing**, often referred to as 'alpha testing', is carried out by other members of your own organisation who have not been directly involved in the development project.
- **Beta testing** comes next. This is performed by your client's users or possibly their customers. It is also known as 'user acceptance testing' and it is often the final stage of testing before the software is formally adopted by the client. The feedback that this generates is essential in confirming that the development was successful.

Other testing considerations

There are other factors and levels of testing that may need to be considered, some of which are covered here.

Environment tiers

Software developers may encounter many different environment tiers while working on a project. The minimum is usually the four shown in Table 10.2, although their names may change depending on the organisation's preference.

Table 10.2 Different environment tiers

Local development	The set-up used by the software developer while the code is being written and debugged prior to committing to version control.
Developmental sandbox	A developmental environment where various types of testing may take place, particularly unit and integration testing.
Staging	A mirror of the production environment which is not public facing but which gives the most complete simulation possible of how the application will work without impacting the end users or the client.
Production	The live environment, which is (eventually) public facing. This has to be correct.

Unfortunately, though not unexpectedly perhaps, issues can occur due to differences between each tier. For example, in a commercial web development project, it is possible that a developer may favour a Microsoft Windows-based local development environment (if allowed to express a preference). This would most likely differ greatly from the live production environment (typically Linux-based) in which their code will eventually execute. This could lead to several problems, which should be tested for in order to uncover issues. For example:

- different behaviours of services – for example, implementations of a web server under different operating systems;
- incompatibilities between different implementations of code libraries – for example, missing or altered functionality;
- case-sensitivity issues in the codebase or database schema (Windows tends to be far more lax than other systems);
- different memory management behaviour at run-time – for example, memory access issues;
- different levels of process support from the operating system at run-time;
- different security settings or access control levels which affect the behaviour of the application;
- different file systems – for example, Windows' NTFS versus Linux's Ext4.

Depending on the context, further testing should also include:

- **Compatibility testing** to test how the software will execute in different environments. Examples include:
    - different versions of a mobile OS, such as Apple iOS or Google Android OS;
    - client-side support of different web browser clients when running a single-page application;
    - different versions of a desktop OS, such as Microsoft Windows and Apple macOS X;
    - different Linux distributions, such as Debian and Fedora.
- **Load testing** is designed to ensure that the system will continue to perform as expected in instances when there is higher demand or when there is a larger number of users. For example, commercial website applications often suffer from scalability issues when demand becomes excessive, principally because this can be difficult to simulate correctly outside the production environment.
- **Performance testing** is where the expected demand on the system is tested in the context where it will generally be used. For instance, it might test how an application behaves under different processing or networking bandwidth constraints (e.g. lower-powered CPUs or broadband vs mobile data).

How much testing is necessary?

Estimating the time that will be needed for testing is not an exact science. Some experienced developers say that software testing should take at least half as long as the time taken for coding, while others say that the amount of time given to testing should be significantly more than half.

The reality is that the amount of testing should be sufficient to rigorously test the final version of the software, prior to it undergoing user acceptance testing (as outlined earlier in this section).

DESIGNING TEST DATA

As discussed earlier in this chapter, white box testing focuses on the structure of the code, so any data that is selected by the developer must reflect a range of values that fully cover the code's logical pathways and branches. As the section above on coverage tests demonstrated, this must include ensuring that modules and functions are tested at least once.

Black box testing is more likely to use real data – that is, the data that will ultimately be a feature of the system.

Test cases

Test cases are used as a tool to help development teams design the right kind of test data for their software. Each test case sets out a number of conditions and/or variables which help the testers to confirm that the software or system is working as it should.

There are many online templates that can be downloaded for use both as individual forms and as test case tables, although many organisations have their own in-house versions. A test case will usually include:

- a test ID;
- the test objective;
- an outline of the procedure needed;
- test data (which should include normal values, unexpected/erroneous values, extreme values and values to specifically test boundaries);
- expected outputs (the expected outcome of the test);
- the test result (which is added after the test has been executed);
- the test status (whether the test passed or failed);
- comments (outlining corrective actions needed).

The test case may also include additional information such as the name of the test, the date of the test and the test environment (such as hardware, software and network details). At the simplest level, what test values could be used to test the following small example of code expressed as pseudocode (see Chapter 11 for more on pseudocode)?

```
START
  INPUT Age
    IF Age >= 60 THEN
         Issue Bus Pass
  ENDIF
END
```

Table 10.3 examines some possibilities.

Table 10.3 Sample test data

Data classification	Value(s)	Explanation
Normal values	69	You would expect this value to generate a bus pass
	31	You would **not** expect this value to generate a bus pass
Boundary values	59, 60, 61	These are values which are just below, on or above the `IF`'s relational operator. **Only** the latter two should issue a bus pass.
Erroneous values	W, p, $, 2.1	These are the wrong data types and although there is no error trapping built in, the developers should know *how* the software will behave in these circumstances.
Extreme values	0, 120, 200	The minimum or maximum values in a data set. In this example, based on age, we might assume 0 (for minimum) and 120 (for maximum). A value of 200 would be extreme *and* invalid (unfortunately humans don't have this lifespan).

Avoiding confirmation bias

Confirmation bias occurs when a person prefers information that supports their view of a situation, irrespective of what the evidence shows. For a software developer this typically manifests in the somewhat subconscious acts of selecting testers, test subjects or test data in such a way to make the testing more likely to produce the desired validation that the code works just as expected.

To help you avoid confirmation bias:

- Choose much more bad test data than good.
- Use others to help you select test data, particularly individuals who have not been involved in the generation of the code.
- Use automation and test tools. These can improve not only the efficiency of testing but also the throughput. Automated testing tools such as TestComplete and Ranorex can contribute to the testing process.

ANALYSING TEST RESULTS

During testing, the results of each test must be carefully recorded. The results are then analysed so that a series of corrective actions can be planned and undertaken to fix problems found.

Corrective actions

Dealing with corrective actions is basically a six-step process:

1. Identify a fault (i.e. a bug or inconsistency in the application's behaviour) through testing.
2. Distinguish the **root cause** of the fault from the symptoms – of course, this may require additional testing and it can be a very time-intensive process to make the symptoms eventually reveal the culprit.
3. Remedy the fault, which may require escalation. For example, a junior developer may need to consult a more senior colleague to obtain advice and guidance before their suggested fix is committed, just in case they make things even worse or break associated code through unintended consequence – something which occurs very often in the commercial sector, unfortunately.
4. Check that the same fault doesn't occur elsewhere (and has escaped discovery somehow). This is particularly important if the problem is tracked to a single developer or a shared library. Identify and fix these also.
5. Repeat the tests to confirm that the fault is no longer present and that the fix hasn't introduced any new issues.
6. Document the changes made (this is important), whether using in-code comments or version control commit messages. Other developers need to know what has been modified.

Preventative actions

A supplementary process is preventative actions. These proactive steps rely on adequate monitoring of application performance during the software development process, typically highlighting and rectifying potential problems before they occur.

Some potential preventative actions include:

- Identifying vulnerable, no longer supported or incompatible software development libraries.
- Engaging with developmental advice regarding improvements in the codebase.
- Identifying patterns in root causes of corrective actions – for example, a developer habitually validating data incorrectly. This may lead to a professional development conversation or some active pair programming (see Chapter 7).
- Integrating improved products or services available to the development team during production. For example, automatic refactoring (see Chapter 7) or linting (see Chapter 12) tools may reduce the likelihood of coding errors occurring.

At the end of the development project, the tests themselves should also be analysed so that the effectiveness of the testing processes can be evaluated.

TIPS FOR TESTING

The following tips may help you in your testing:

- Be organised and prepare – testing is an important part of the software development process.
- Ensure that sufficient time has been allocated to the activity.
- Test often and early – fixing errors earlier in the development cycle is cheaper.
- Ensure coverage is as close to 100% as possible.
- Don't forget the need for robustness in a solution – malicious attacks, particularly from attack vectors such as unvalidated user input, are becoming more and more likely.
- Integration testing may highlight compatibility issues between software components (units, modules etc.).
- Test cases and the choice of test data should not be overly complicated and you should be able to justify your test data choices.
- Testing should be carefully documented so that its effectiveness can be evaluated alongside all other aspects of the development project.
- Tests should be prioritised, particularly where there are time constraints, to focus activity.
- Don't forget to involve the end users as they will also produce testing output.
- Don't forget to test the end user and technical documentation.
- Avoid confirmation bias – all developers suffer from it at some point.
- Don't be afraid to escalate an issue to a more senior developer if you are not sure – collective responsibility should always be encouraged within a development team.
- The aim of testing is to ensure that the product is stable and fully meets the client's requirements.

SUMMARY

In this chapter we have explored some ideas and concepts around typical testing activities. The contribution that thorough testing makes to a successful software development project cannot be overestimated. Development teams increasingly employ software production pipelines which prioritise comprehensive white box and black box software testing. Bringing these forward from the later parts of the process enables faults and bugs to be caught much earlier, which means they take less time to fix. This makes the whole creation process a lot cheaper. Just for once, everyone can win.

The next chapter explores structured techniques used in problem-solving in software development.

11 WORKING WITH STRUCTURED TECHNIQUES TO PROBLEM-SOLVE AND DESIGN SOLUTIONS

'If you define the problem correctly, you almost have the solution.'

– Steve Jobs

Generally, solutions don't design themselves, although advances in machine learning and artificial intelligence are building a convincing argument that things in this regard are slowly changing. For most developers, solutions still rely on human ingenuity and traditional problem-solving methods.

Designing and communicating solutions in a standardised manner is a good way to discuss and debate ideas with other members of a development team. What role does documentation play? And should you ignore analysis and dive straight to the keyboard and code a solution?

In this chapter we'll explore various techniques which can help you to organise your thoughts and communicate them with your peers.

DESIGNING AND RESOURCING THE SOLUTION

Designing a solution is a complex process, and many structured techniques can be applied to help you design effective solutions.

Abstraction

Abstraction is a useful process which aims to filter out irrelevant details from the solution; basic information is required but the specifics are, in the first instance, not important. For example, we know a postcode needs to be validated – so there will be a discrete module which takes a postcode and says whether it is or is not valid. Just how that works at the moment isn't important – **we abstract that away for now** (it's just a specific implementation detail) and deal with it later in the implementation.

Modularity

Whether you are working on a large or a small project, it is likely that both the problem and the development will be broken down into manageable chunks of activity, often focusing on a single function or feature. This is known as modularity.

A modular approach makes sense whether you are a single programmer working on the development of a website for a small business or a member of a development team working, for example, on a new patient management system for the NHS where you may only ever be involved in delivering a small part of the larger system.

Decomposing the problem into discrete activities has many benefits. It helps you to conquer complex problems which can otherwise feel overwhelming and it provides the opportunity to reuse code extracts and modules in different areas of the solution. In addition, it makes the maintenance of the software easier, because to resolve an issue you will be working with fewer lines of code. And, importantly, modularity promotes an incremental approach to development where parts of the system can be delivered in stages.

One downside of a modular approach is that you may fail to benefit from optimisations that could otherwise have surfaced (e.g. code that is separated into a module may escape detailed review and therefore potential improvements which may have been identified by a more senior developer). Other disadvantages include the potential for reduced speed if too many modules are incorporated and the potential for code to be produced which is ultimately difficult to read. There is also the potential that bugs from third-party code may be introduced.

But there are other considerations when it comes to the design and resourcing of a development activity.

Elasticity of resources

'Elasticity of resources' is a term drawn from business. The term 'elasticity' is used to define the extent to which a particular resource can be stretched, adapted or manipulated in order to accommodate the activities or needs of the development project.

As we know, time is finite. Twenty-four hours cannot suddenly become twenty-eight. However, additional time can be generated in development simply by increasing the number of people working on a project. This may also mean additional computer resources or software will be required to accommodate the extra personnel, which will affect costs (including overhead), and there may additionally be a need for more communication. Some professionals would even argue in favour of Brooks' law, which suggests that increasing the number of people working on a project will simply delay the delivery of the project incrementally in proportion to the number of people who are added because it can take time for the additional staff to become productive.

To make team members more flexible, they may need to upskill (improve their existing skills or acquire new ones), attend courses or programming boot camps, or be given time to explore online learning opportunities. For example, the Python Community (https://www.python.org/community) gives developers an opportunity to discuss ideas and approaches with other experienced professionals. Many of the pages on this site feature articles, case studies, tutorials and Q&A content.

The development budget is almost always the least elastic of the required resources. Clients will rarely agree to further expenditure, unless they can see a definite business benefit in relation to the additional outlay. This will often be linked to the development of additional features or functionality requested during client meetings. In essence, the development scope will often be seen as elastic while the time and budget remain inelastic.

THE IMPORTANCE OF DIAGRAMMING

A well-structured and clear diagram is much better at demonstrating a concept, plan or schema than paragraphs of text. A diagram conveys information more quickly and more directly, and is a more flexible tool than a verbal communication as it also provides a documented record of what was planned and agreed. Diagrams can, after all, be annotated.

To produce appropriate diagrams, developers should think about:

- **The purpose of the diagram:** what message is the developer trying to convey?
- **The audience and their needs**: simplified diagrams will be necessary for a non-technical audience while complex diagrams will suit the more technical.
- **Level of complexity:** diagrams can be created at top level, which may be sufficient to outline the basics of a concept or plan for the client, while more detailed versions may be produced for the development team.
- **Format:** diagrams can be static, simply demonstrating the relationship between different functions or data items, or dynamic, showing how aspects of the solution will change over time.

Most methodologies make use of diagrams as a development tool, to capture one or more ideas for potential solutions to the problem. In the IT industry, many diagrams use the same basic techniques even if they may be visually a little different. Each diagram has a different use. Some examples of appropriate development diagrams follow, together with examples of how each diagram would contribute to an understanding of the problem and the plan for the solution.

Entity relationship diagrams

Entity relationship diagrams (ERDs) are used to articulate the relationship between the data items in a database. They are used to represent the current state of the data during the problem analysis and to articulate possible approaches to solving the problem.

Each of the boxes in the diagram is known as an entity. It is an object about which data is stored. Consider the ERD in Figure 11.1.

Figure 11.1 A simple entity relationship diagram

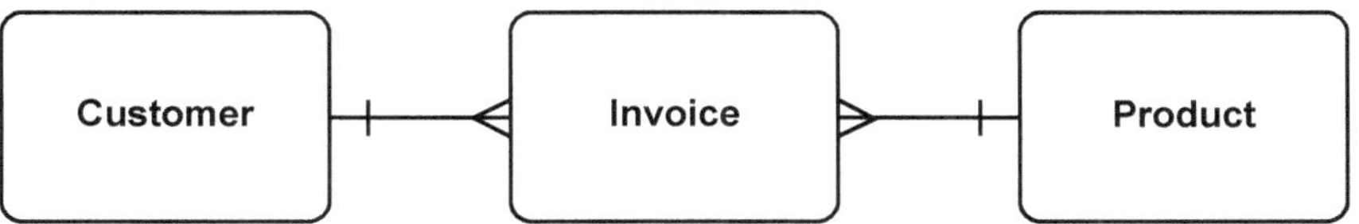

This is a visual representation of the data that will be needed to produce a customer invoice. The lines between the entities explain the relationship. The diagram means:

- One Customer can have many Invoices: one-to-many ─┼──<
- One Product can be on many Invoices: one-to-many in the opposite direction >──┼─

However, when you read the diagram, you need to read the relationships from both sides. Figure 11.1 would be read as:

- One Customer can have many Invoices and there will be many Invoices for one Customer.
- One Product can be on many Invoices but many Invoices will be for one Product.

When you read it from this perspective, the reality is that the diagram is flawed – particularly the relationship between the Invoice and the Product, because most Invoices will contain many Products. Therefore, the actual relationship between an Invoice and a Product is many-to-many (see Figure 11.2), and a many-to-many relationship should not exist in a well-designed database.

Figure 11.2 Revision of the entity relationship diagram in Figure 11.1

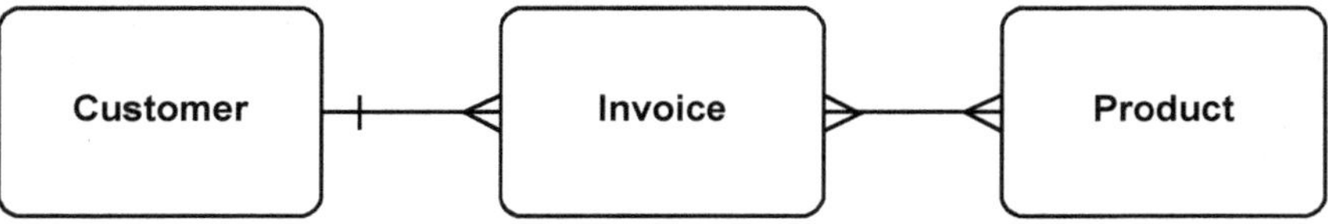

To resolve the relationship between the Invoice and the Product, we need to further define it (as shown in Figure 11.3):

- One Customer can have many Invoices and there will be many Invoices for one Customer.
- One Invoice will have many Invoice Lines and there will be many Invoice Lines on one Invoice.
- One Product can be on many Invoice Lines but the many Invoice Lines will be for one Product each.

Figure 11.3 Correction of the entity relationship diagram in Figure 11.2

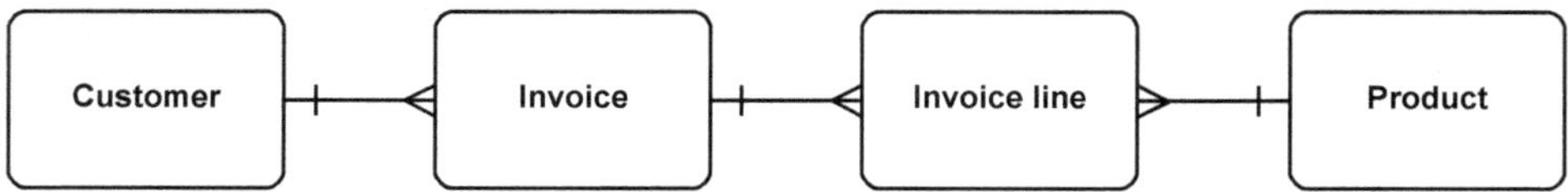

Having identified the entities and the relationships, the next step would be to identify all of the data items for each entity. Figure 11.3 would need to be supported by a list of data items which the developers would use to create the system. The alternative is another version of the diagram as shown in Figure 11.4, where the data items are included in the diagram.

Figure 11.4 A detailed entity relationship diagram

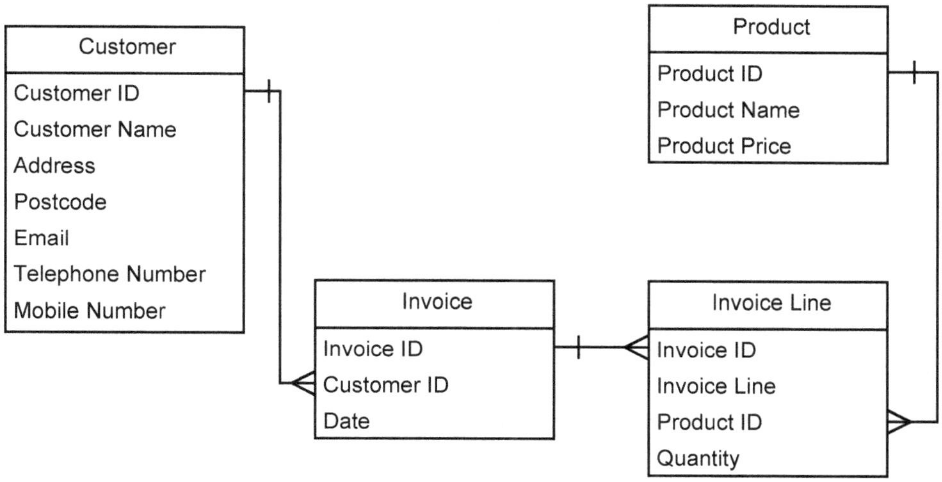

Customer
Customer ID
Customer Name
Address
Postcode
Email
Telephone Number
Mobile Number

Product
Product ID
Product Name
Product Price

Invoice
Invoice ID
Customer ID
Date

Invoice Line
Invoice ID
Invoice Line
Product ID
Quantity

In addition to the diagram, there would be a data table which would list all of the data items, data types and sizes (in the case of text), as well as any default values (such as 0 for quantity), any validation being applied and any formatting decisions (such as the formats of dates, currency or postcodes).

Data flow diagrams

A data flow diagram (DFD) represents the logic of how data flows through a system. It articulates how the data is triggered and what the processes are that use or change this data.

Figure 11.5 expands on the invoicing system concept and shows what a DFD might look like. Everything inside the larger box represents activities which contribute to the sale (i.e. invoice generation). Items 5, 6 and 7 are the data files suggested in the ERD. Although in the final ERD there were four tables, Invoice and Invoice Line exist on the same document. You should be able to see how these documents fit together to produce a view of the system.

Figure 11.5 A data flow diagram for an invoicing system

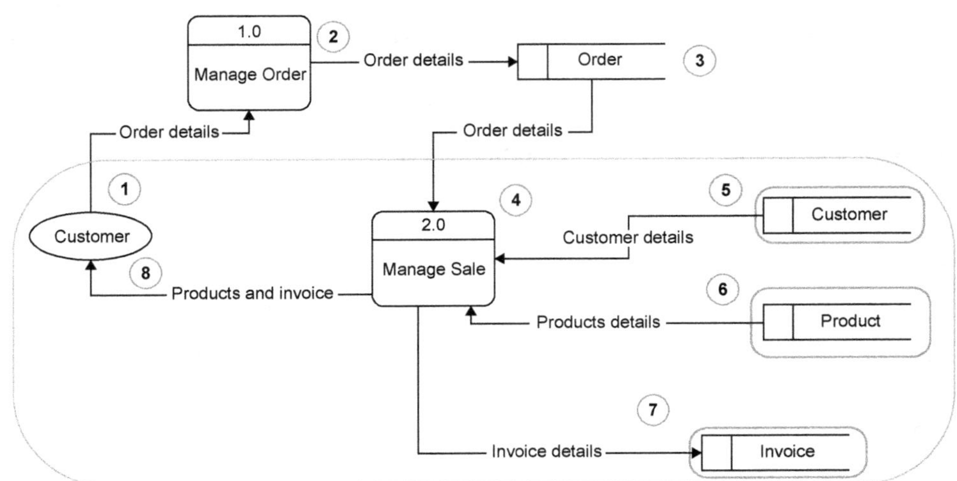

You will notice:

1. The customer order triggers the activity.
2. An order is created.
3. The order is stored (the order is not included in the ERDs shown in Figures 11.3 and 11.4, but it would be in an extended version of the diagrams in a real development).
4. The order is then used by the sales process to make the sale.
5. The customer details are accessed.
6. The product details are accessed.
7. An invoice is stored.
8. The products and the invoice are sent to the customer.

Flowcharts

Used alongside DFDs, flowcharts explain the processes that are coded to change the state of data or to create new data in a project. In our example, the generation of an invoice would occur as shown in Figure 11.6.

Although a very old technique, flowcharting remains a popular one used by developers to work out the flow of an algorithmic process and communicate it visually to others. It is an instance of a picture painting a thousand words.

There are many other diagram types used in different contexts, such as wireframe diagrams used in website and mobile app design. The wireframe diagram in Figure 11.7 demonstrates navigation between mobile screens in a proposed mobile app.

Figure 11.6 Invoicing flowchart

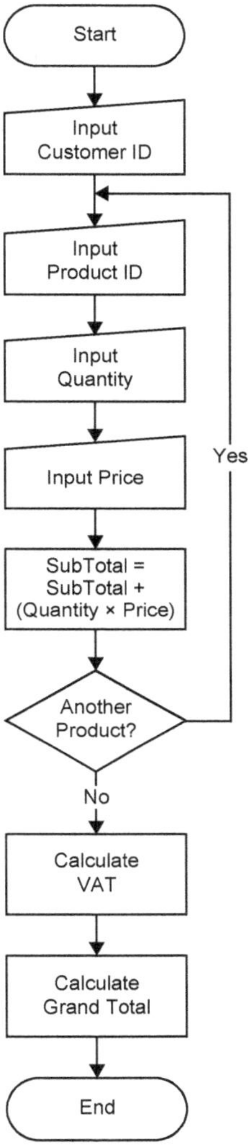

The diagram starts and ends with a terminator symbol.

The user inputs the Customer ID, denoted by the **manual input** symbol.

The user inputs the Product ID followed by the Quantity and then the Price.

The Quantity is multiplied by the Price for this item and is added to the running SubTotal. (This SubTotal is **calculated** pre-VAT.)

There is a **decision** box in the flowchart; the user is asked whether another Product is to be processed. If so, the program loops to get another Product ID, Quantity and Price. If there are no further Products to be input, the program continues to the calculation steps.

The VAT is calculated using the running SubTotal.

The invoice total is calculated.

The process is ended.

Notice that Figure 11.7 represents the content and possible positions of components. It does not have any colour choices, preferred fonts or branding at this stage.

There is more on diagramming in books that support specific methodologies. But, for the practical application of diagramming techniques, you will find some useful videos on YouTube (see SmartDraw 2017, 2018a, 2018b).

Figure 11.7 A typical wireframe diagram

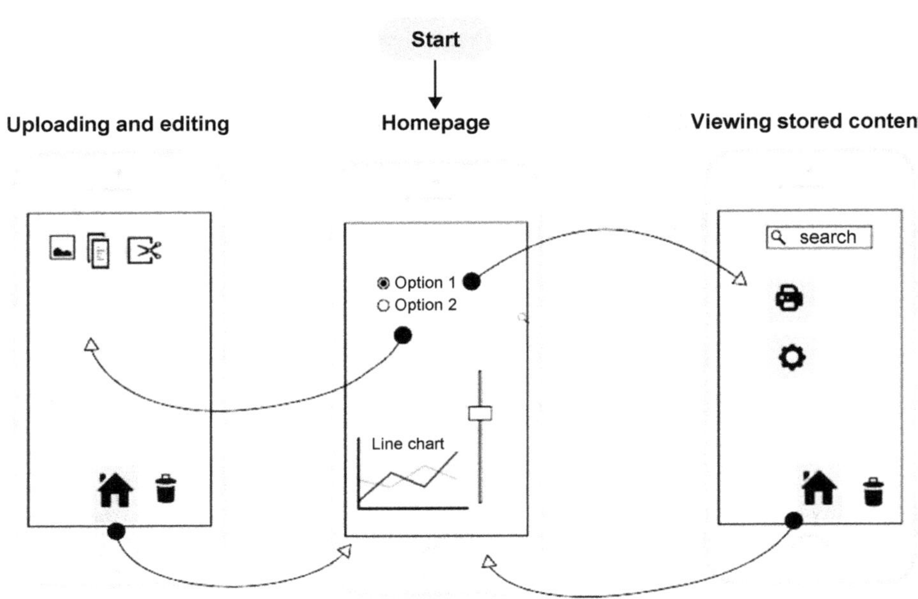

Unified Modelling Language diagrams

Unified Modelling Language (UML) is a standardised set of diagrams often used by software developers. UML includes:

- use case diagrams;
- activity diagrams;
- object diagrams;
- class diagrams;
- component diagrams;
- sequence diagrams.

As an example of an UML diagram, a use case diagram for the invoicing system would look something like the content of Figure 11.8.

The central ovals in the diagram represent the system while the stick men, known as actors, represent entities outside the system that the system interacts with. Inside the system are four processes: products purchased (which go to the customer), payment of invoices (coming in from the customer), payment of invoices (going out to the supplier) and products coming into the system (from the supplier).

Figure 11.8 A Unified Modelling Language use case diagram for an invoicing system

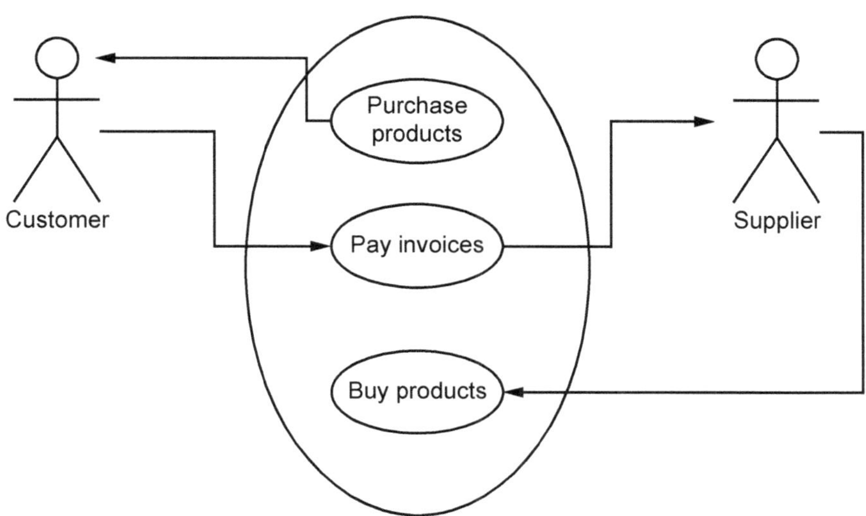

There are many similarities between this diagram and a combination of a DFD, a simplified flowchart and an ERD. In the real world you will find that each organisation will have its own organisational preferences in terms of which diagram types it uses in its development activities. Think about diagrams as a language all on their own. It might take time to learn to 'speak' the language, but often, even with little knowledge, you can understand what the diagram represents. The main thing to remember is that as a developer you will need to choose the diagramming tools you use to suit the context of the development activity at hand.

As a developer it would be beneficial for you to explore some other UML diagrams as these will be useful in your development practice from time to time. Books such as *UML 2 for Dummies* (Chonoles and Schardt 2013) provide a good introduction for beginners.

Pseudocode

Pseudocode is essentially a line-by-line description of each action needed in a program. Sometimes it is used as an alternative to flowcharts to represent the flow of a proposed program. On other occasions it is used alongside flowcharts.

Each step is defined in the order that it will occur in the program. Key words such as **START** and **END** show the start and end of the program, or of a function or module. **REPEAT UNTIL** indicates a loop will be needed, and **INPUTS**, **CALCULATIONS** and **OUTPUTS** are identified (e.g. a calculation may be required before an output).

Here is a pseudocode version of the flowchart in Figure 11.6:

```
START
  Input Customer ID
  REPEAT
    INPUT Product ID
    INPUT Quantity
  UNTIL Another Product = 'N'
  CALCULATE Total
  CALCULATE VAT
  CALCULATE Grand Total
END
```

The following example demonstrates how pseudocode handles selections – in this example, an **IF** statement:

```
START
  INPUT Age
  IF Age >= 60 THEN
     Issue Bus Pass
  ENDIF
END
```

The equivalent flowchart is shown in Figure 11.9.

Figure 11.9 Bus pass generator flowchart to create one bus pass each time the program executes

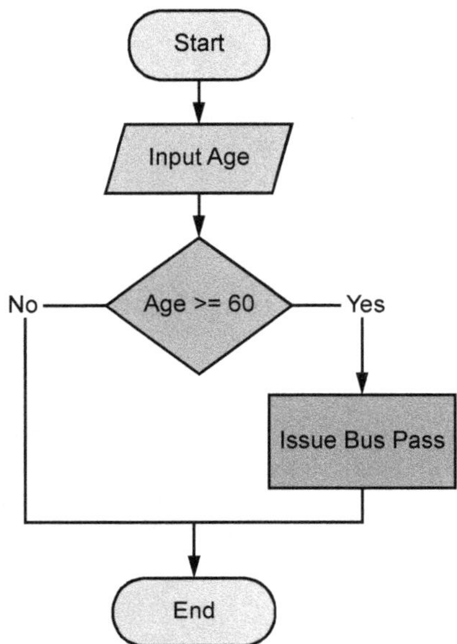

You will note here that there is no **ELSE** as there are no **ELSE** actions to accommodate. The **ELSE** would have been an alternative action if the condition had not been met. In this case there are no actions to perform. Either the person is aged 60 or more, in which case the bus pass is generated, or they are not aged 60 or more, and there is no counter-action to perform.

In the following adapted version the user is asked, after each pass through the code, whether another pass is needed. This example includes both a selection and a loop. The equivalent flowchart is shown in Figure 11.10.

Figure 11.10 Bus pass generator flowchart to create multiple bus passes

```
START
  REPEAT
    INPUT Age
    IF Age >= 60 THEN
      Issue Bus Pass
    ENDIF
  UNTIL Another Bus Pass = 'N'
END
```

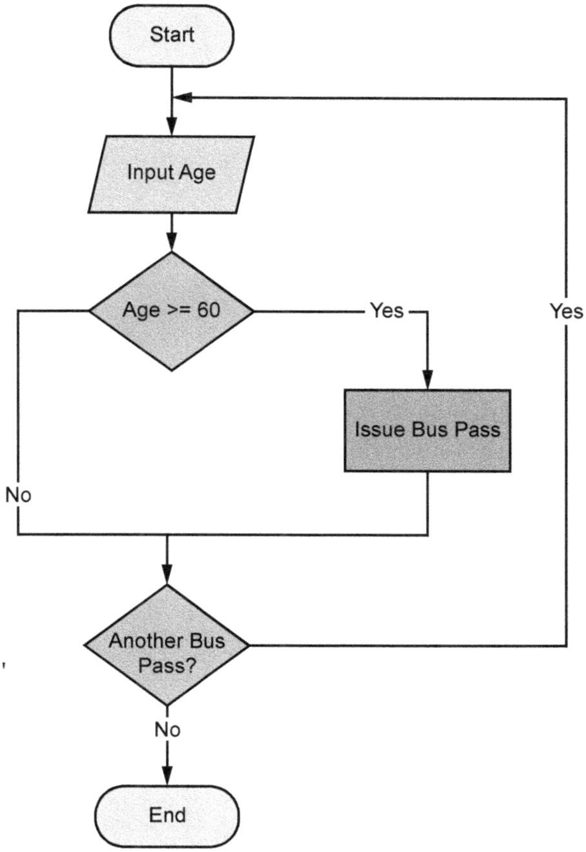

Pseudocode has no formal standard, unlike some of the diagrams you have seen in this chapter of the book. Having said that, an organisation may well have a preferred house style.

Decision tables

Decision tables are another useful tool for capturing requirements. They eventually form the core of a solution's logical testing and often help to crystallise the rules involved when modelling the business decisions for a software project.

A decision table for a selected process is split into three parts:

- **conditions:** key questions or statements of fact that must resolve as Boolean values (i.e. true/false, yes/no etc.);
- **actions:** each action that can be performed in the process;
- **rules:** formed from a combination of the conditions and actions.

Table 11.1 shows a practical example of a typical set of conditions for a traditional credential-based user login.

Table 11.1 A decision table for testing a user login process (X indicates an action)

	Rules					
Conditions	**R1**	**R2**	**R3**	**R4**	**R5**	**R6**
C1 – Username correct	N	N	Y	Y	Y	Y
C2 – Password correct	N	N	N	N	Y	Y
C3 – Password has expired	N	Y	N	Y	N	Y
Actions						
A1 - Reject login	X	X	X	X		
A2 - Accept login					X	X
A3 - Force password change						X

In order for the user to log in, **both** credentials (username and password) must be correct. And, even if the credentials are correct, if the user has an expired password (e.g. if the system requires passwords to be changed every three months), the program should start an additional process which forces the user to select a new one.

Remember, from a security perspective, you should never separately acknowledge that a username is correct – that's an open invitation for brute-force hacking, social engineering and potential escalation.

All of this descriptive business logic can be represented in a simple decision table that clearly indicates which conditions must be true for the various actions to be performed. When the business logic becomes more complex and convoluted, this type of visual representation helps designers and programmers to communicate their ideas quickly and clearly.

CONFIRMING THE DESIGN WITH THE CLIENT BEFORE CODING

All design decisions should be shared with the client before the final, working version of the design documentation is put together.

A formal presentation to the client and any other relevant stakeholders will usually be undertaken by the project manager and some of the key development staff, who will be available to answer any technical questions if required. The presentation will be used as the basis for discussion, to clarify understanding on both sides, to reach a consensus and to shape the formal design documentation.

Design documentation

Design documentation should be sufficiently detailed to enable a development team to produce the solution. It should include:

- an overview of the product, as agreed and signed off by the client – this will help to remind the development team what has been promised to the client and may reduce the possibility of scope creep;
- a full breakdown of expected milestones and deadlines;
- a list of user requirements including all functions and features expected;
- documentation which describes the expectations for the user interface;
- data design documentation which describes the data structures and data objects needed;
- appropriate diagrams to explain key concepts and demonstrate the flow of the relevant processes.

Legal implications

There are significant laws as well as ethical and social factors that should be considered by a development team in the creation of a software product.

Discrimination law, such as the Equality Act 2010 (which replaced the Disability Discrimination Act 1995), protects individuals from being treated unfairly based on their personal characteristics or life choices. One area of the act covers goods and services provided by organisations, which should strive to be inclusive and accessible to all groups. You are therefore expected to design your products with accessibility features which will allow a wide variety of users to access the product features, sometimes using different devices or additional software such as screen readers. The software your team creates should enable a disabled user to use the software having made 'reasonable adjustments' (Equality Act 2010).

For this reason, the design of software and websites should be influenced by the needs of a range of both able and disabled users, such as those with blindness or visual impairment, motor impairment and cognitive impairment. Additionally, in the case of websites in particular, your design should conform to Web Content Accessibility Guidelines 2.0, published by the World Wide Web Consortium (see https://www.w3.org/TR/WCAG20).

Developers should also be aware of the notion of intellectual theft. This is where another software product's features are intentionally copied or digital assets such as logos, symbols, names, illustrations or photographic images published elsewhere on the internet are used without the permission of the owner of that intellectual property. There may be an opportunity to use some of these assets for payment, under licence or subject to a royalty. In any case, you need to be sure that you have permission to use anything created by anyone outside your development team. The Intellectual Property Office produces copyright notices that cover digital images and photographs and the regulations on their use (see Intellectual Property Office 2016).

As a developer you should become familiar with key copyright terms including 'commercial', 'shareware', 'open source' and 'freeware'. All of these are still covered by copyright but use is permitted in specific contexts. In contrast, when something is in the public domain, it means that copyright has been relinquished by the copyright owner (or has expired).

From a software development perspective, the development team has a range of responsibilities in addition to those already outlined above, including:

- demonstrating a consideration of ergonomic issues in the design of software;
- designing software that does not generate or transmit any type of malware;
- ensuring that software that uses a network does not compromise the network in any way (such as impacting other services sharing the same network);
- ensuring data privacy and security are not compromised.

In other words, every effort should be made to ensure that code is robust and that there are no flaws or loopholes that will compromise security in any way. In essence, this means programming for cybersecurity. See *Cyber Security: A Practitioner's Guide* (Sutton 2019) for an overview of different threat types and vulnerabilities.

Complying with the GDPR

The other key legal implication is compliance with the General Data Protection Regulation 2018 (referred to as the GDPR). This superseded the Data Protection Act 1998, which had failed to keep up with increasingly swift technological change.

The GDPR is probably the most important legislation which affects IT and data use in the UK, and the penalties are high for organisations that fail to comply. Breaches of the GDPR could result in a fine of up to 4% of the organisation's turnover or a penalty of €20 million, whichever is the greater.

The GDPR applies to every organisation in the world that processes the personal data of the citizens of the European Union (EU). Because it is an EU regulation, you may think that the UK's withdrawal from the EU following Brexit would have made this legislation irrelevant. This is not the case because even if the UK decides not to retain the GDPR for citizens of the UK, trading with any member state of the EU will require UK businesses to comply with the terms of the GDPR. You may be aware that this has already affected businesses around the world that trade with the EU nations; some such businesses have already chosen to cease trading to avoid the need to comply with the legislation.

The best way to help your client to comply with the GDPR is to design and build a system which conforms to the requirements of the GDPR from the outset. This will demonstrate an element of quality and will enhance the client's trust of the product. As a developer it is your responsibility to educate yourself in relation to the full requirements of the GDPR.

Any data that can be linked to an individual falls under the GDPR, including:

- name, address, telephone number and email;
- date of birth, gender, religious beliefs and sexual orientation;
- photographs;
- medical history;
- posts to social media;
- IP addresses;
- financial information.

These details are categorised as personal because they can be used to establish an individual's identity. You should also be aware that if an organisation wishes to process data belonging to children aged 16 and below, it has to seek parental consent.

The main principles of this legislation require that:

- Personal data must be processed **lawfully, fairly and transparently**. This means that the organisation must prove that it had a valid reason for obtaining and using the data. The processing of data must in no way mislead, and data must never be used in a way that is detrimental to the data subject. In addition, the organisation must be honest with the data subject in relation to how the data will be used.
- Personal data must be **collected for a specific purpose** and it must not be further processed to be used in other ways. For many years, organisations had been selling data (usually in the form of mailing lists) to third parties, particularly for use in marketing activity. The GDPR has made this practice illegal and this has led to a reduction in spam texts and emails.
- The personal data that is collected must be **limited to what is actually needed** for the purpose for which it is being collected – organisations are no longer allowed to gather and store more data than they need for legitimate purposes. For example, does an organisation need your date of birth? If not, then they may not collect and store it.
- Personal data must be **accurate and up to date** – it can be very annoying if your name is not updated on a system after a change.

- **Data must not be stored for longer than is necessary** unless storing is in the public interest or it is to be used for scientific research purposes. For example, hospitals have to report their activities, producing statistical data about the number of operations carried out, conditions treated, illnesses examined, accident and emergency waiting times, and so on. Most of us would agree to our data being used for these purposes, but in order to be used the data has to be anonymised to remove personal information such as names and addresses.
- Personal data must be processed and stored in a way that is **confidential and secure** to ensure that individuals are protected. Data privacy and security should be at the top of the developer's agenda and everything should be done to ensure that the data is not compromised, including ensuring that the data is safe from hackers, is backed up securely, and is erased if the data subject invokes the 'right to be forgotten', meaning that they have made a request for their data to be removed from the system. From a software development perspective, this may require you to develop some automated processes to cleanse data from your client's system.
- The final principle of the GDPR concerns **data transfer**: data must never be transferred to another country which does not have its own suitable and sufficient data protection legislation. The reality is that businesses in the UK may not even share data with their own offices in other countries if those countries do not abide by the requirements of the GDPR.

Under the GDPR, individuals have the following rights in relation to their data:

- the right to access the data being stored about them;
- the right to ask for information about how their data is being processed and used, and the right to be allowed to withdraw their consent to these activities;
- the right to ask for their data to be erased (known as the right to be forgotten);
- the right to request their data and reuse it.

With the high penalties that may be invoked if organisations breach the GDPR, most businesses take it very seriously indeed.

How not to over-promise and under-deliver

In an attempt to win a contract, many organisations cut their costs and promise more than their competitors. Offering a bigger and better product cheaper, however, does have implications for the organisation.

If you have had to undercut competitors, you may no longer be able to afford the team you originally selected to work on the potential project. This may mean that some of the developers will need to work on other income-generating projects instead, which can have a detrimental impact on the development project:

- You may lose the breadth of skills needed.
- Fewer people will be available but they will still be expected to manage the same workload as would have existed for a larger team.
- The team will still be expected to meet the same deadlines but with fewer people.

- You may actually run out of time, and in some contracts there may even be a 'late clause' with financial implications.
- The general quality of the product will be compromised.
- There may be a detrimental impact on the client's business. There are many well-documented examples of software bugs which have been missed due to insufficient testing and which have negatively affected a business' brand. Managing bugs requires significant activity (see Chapters 10 and 12) and affects the maintenance of the product going forward.

The most serious impact is damage to your organisation's reputation. This can be hard to come back from. Following are some tips to help you avoid falling into the trap of over-promising and under-delivering:

- If you do have to work with a smaller team, you should ensure that the focus is on producing a quality product even if this ultimately reduces your profit margin to a degree.
- You will have to undertake close monitoring across the entire lifecycle to ensure that a quality product is produced. This will enable you to take positive action earlier if you think that quality is being compromised.
- You could consider outsourcing elements of the development project, such as testing, to third parties, particularly in other countries where labour may be cheaper. However, this should only reduce your in-house testing activity, not remove it completely.
- Consider a strategy of continuous testing throughout development. Any issues arising will be uncovered much more quickly.
- Use technology where you can, such as test management tools, to improve the efficiency of your testing strategy and reduce your overall costs.
- Choose a development methodology such as Agile which uses the concept of transparency as one of its cornerstones to reduce risks around quality, cost and time. Transparency implies a degree of openness within the team, built on a foundation of communication, accountability and honesty. It enables development teams to make better-informed decisions.
- Be realistic about what the development team can achieve. Sometimes this requires you to be as honest with yourself as you are with your client.
- Don't be pressured into exceeding expectations.
- Ensure that all agreed points are documented and signed off by the client to avoid disagreements further down the line and potential legal action.
- Think carefully before you agree to any contract. As hard as this may seem, be prepared to walk away from a project if necessary. Producing a sub-standard product is in nobody's best interests and can result in reputational damage for all parties (and much worse, including financial wrangling and litigation).

SUMMARY

In this chapter we have discussed how you can work with structured techniques and have given you an overview of how to avoid some of the common pitfalls experienced by novice developers.

However, even with the best possible planning, multiple issues will arise between the interpretation of solution designs and their eventual implementation. When this happens, it is important to be able to understand the underlying program structure (in comparison to its design) and use appropriate tools and techniques to debug the code; we'll examine these in the next chapter.

12 HOW TO DEBUG CODE AND UNDERSTAND UNDERLYING PROGRAM STRUCTURE

> 'What kind of programmer is so divorced from reality that she thinks she'll get complex software right the first time?'
>
> – James Alan Gardner

Debugging is the process of locating an error in program code and fixing it. It's often not a fun process, it can take a disproportionate amount of time and it's often caused by a very silly mistake (rather than something technically deep). Despite this, debugging is a critical activity and developers need to use the tools available to them to make the process as painless as possible.

Failure to debug will ultimately result in insecure code and dissatisfied clients and users. There are financial implications to this as well as reputational ones, as discussed in Chapter 11.

WHEN SHOULD DEBUGGING OCCUR?

That's a surprisingly good question. A large amount of debugging actually occurs **before** the program code is formally tested (i.e. it's done during the coding phase). This is because developers often get the syntax of a command or function wrong when they are writing code. In attempting to informally test their snippet of code in their local environment (to check their progress), they instead generate a fatal 'syntax error'. This essentially means that the developer has not abided by the rules of the target programming language and its translator (interpreter, compiler or assembler) cannot convert the source code written to executable code for the CPU to run.

Syntax errors are generally easy to solve because modern programming languages tend to supply informative error descriptions (this wasn't always the case). Here's a short Python example to demonstrate:

```
cel = 100
fah = c * 9/5 + 32
print(Fah)
```

```
Traceback (most recent call last):
 File "C:/Users/Test/mytemp.py", line 3, in <module>
  print(Fah)
NameError: name 'Fah' is not defined
```

In this case, the resolution is fairly straightforward – we've tried to output a variable called **`Fah`** rather than the variable **`fah`** (Python, like most languages, is case sensitive). The **`NameError`** tells us the problem and helpfully also supplies the line number (**3**). If we fix the variable name, the code works fine.

Of course, not all syntax errors are as easy to fix!

Generally speaking, the ability to debug errors improves with experiential learning – the more you successfully solve and the more diverse the errors are, the better you eventually become.

A useful (and well-known) tip is to ask another developer to take a look if you're stuck; they will approach the code (and the error) with a fresh pair of eyes and will often identify the offending issue quite quickly. In a fast-paced coding environment, it's not a good use of time to keep banging your head metaphorically against the wall for too long. Learn to escalate if you need assistance – and this is true for developers at any level.

DEBUGGING TOOLS

To a degree, debugging tools depend on the language used and the selected toolchain. Most modern integrated development environments (IDEs) have comprehensive debugging tools.

Common debugging tools include:

- **automatic code completion:** suggests possible alternatives to finish a statement or function;
- **automatic code correction:** does what it says on the tin;
- **embedded linting:** checks for common coding problems (usually stylistic), such as badly chosen variable names, overlong lines of code, inconsistent logic and unused variables;
- **breakpoint:** pauses execution at a given line number, allowing the state of the CPU or RAM (for example) to be inspected;
- **watch or inspect:** allows the developer to 'see' the current contents of a variable (or block of RAM) once a program is paused;
- **trace into:** allows the developer to execute code line by line and observe the execution flow (this includes tracing 'into' a function from its call);
- **step over:** allows the user to execute a function as a black box while tracing;
- **help facility:** provides technical documentation for the syntax of the programming language, often integrated into the code completion/correction functions of the IDE.

COMMON DEBUGGING TACTICS USED BY DEVELOPERS

There are a number of tactics often used by developers when debugging their code. It should be noted that not all of them are advisable:

- **Using debug messages:** developers may use 'print' statements or 'dump' mechanisms to display messages or the content of variables at certain parts of the program. Their aim is to confirm the actual execution flow and the content of the variables at that point against their expectations. This can be particularly heinous if the debug message somehow manages to become merged into the live codebase. Clients don't want to see your debug messages – it's unprofessional and reflects very poorly on the development team.

- **Strategic 'commenting-out' of program code:** suspicious code which 'smells' may be commented out to see what effect it has on the overall solution. This process is typically refined until the offending line of code is identified – this is a common practice when tracking a semantic error.
- **Watching the error log in real time:** when programs are running, they often generate warnings which do not cause a crash. Use of a 'tail' command allows a developer to view new additions to an error log file as the application executes, often enabling them to cross-reference the cause of the problem with its associated entry in the log.
- **Follow the stacktrace:** when errors are displayed, they often show a stacktrace (a report showing the active stackframes and their function calls, frequently shown in reverse order from the error back to the initial call). Following this chain level by level often helps the developer to understand the flow of data between function calls and where potential errors might be.

SEMANTIC ERRORS

A semantic error is usually caused by incorrect program logic (an error in 'meaning'). Because the error is **not syntactical** in nature, the translator will unfortunately not complain (although a linter might...).

For example, here is a C statement that is intended to check whether a user is over 18 years old (i.e. not a minor) but actually contains an error:

```
//check user is not a minor…
If (user_age < 18) {
```

This incorrect use of the **<** operator (essentially a silly typo) would not be flagged by the translator. However, it would unexpectedly change the execution flow of the program when run – not good!

Even if this semantic error were not caught during development, it should certainly be caught during formal testing, probably leading to a few red faces.

Typical semantic errors include:

- incorrectly written arithmetic calculations (usually an incorrect order of operations; see Chapter 5);
- incorrectly written relational and/or logical expressions, for example using **<** rather than **>**, **`and`** instead of **`or`** etc.;
- incorrect use of data types, for example a string rather than an integer or vice versa;

- incorrect placement of brackets, impacting the desired order of operations;
- incorrect order of coded statements;
- uninitialised variables being used in calculations.

Semantic errors are often much harder to debug correctly and can be very time-consuming to resolve.

Tools such as linters can be used to identify sylistic errors, declared (but unused) variables, suspicious constructs, incompatible data types, etc. Linting functionality is often built into modern IDEs.

MAKING THE UNDERLYING PROGRAM STRUCTURE MORE OBVIOUS

Debugging code is difficult when attention hasn't been paid to the underlying structure of the program code. It can sometimes be impossible to identify where a construct (such as a loop or an if statement) begins and ends. This is why, as mentioned in Chapter 7, indentation and commenting are so important. Put simply, if it's difficult to read, it's difficult to understand and therefore is difficult to test effectively.

Although program code may be conceptually hard to understand at times (even for the most experienced developers), there's no reason it should be difficult to read. Most programming languages use indentation schemes to highlight the logic involved in a solution – for example, in selections and iterations.

Developers can use many other tactics to highlight the structure of their solutions, such as:

- writing code in concise, separate modules (functions, procedures, subroutines etc.) which follow the single responsibility principle – each has a single, well-defined job;
- ensuring all identifiers (modules, constants, variables, custom types, classes etc.) are meaningfully named;
- creating variables which have limited scope (which means the variable can only be accessed by the function, procedure or subroutine in which it occurs);
- adding real-world comments (where appropriate – too many can be as confusing as too few);
- writing complex code in such a way that it is not unduly complicated in approach;
- making use of existing functionality rather than unnecessarily reinventing the wheel (poorly).

Above all, remember that maintenance is often carried out by another developer. Ask yourself: is my code readily understandable by others?

TIPS FOR DEBUGGING

You may find the following list of debugging tips useful to help you to plan debugging activity:

- Be methodical and check the error message carefully – what is it **really** saying?
- Remember, although a line of code may be flagged as the culprit, the underlying cause may be earlier – an incorrect calculation, bad assignment or erroneous expression are all possibilities.
- If the error has only recently been introduced, review what has been added, modified or deleted.
- Warnings may indicate potential run-time errors (particularly in C-family languages) and should always be resolved.
- Don't forget to leverage the full range of debugging tools available in your chosen toolchain.
- Careful use of breakpoints, tracing and watches often helps to track down semantic errors.
- Debugging improves with practice, and always offer to help others.
- If you really can't see the error, ask a colleague for help but make sure they explain the problem, how they found it **and** how they fixed it.
- Correct indentation and meaningful comments often help other developers to understand code a little better. This always assists the debugging process.

SUMMARY

No matter how carefully you code, bugs **will** occur – it's inevitable. Writing your code using principles that reveal the underlying structure of the code will assist debugging. Remember that while modern tools will identify syntax errors and can identify odd coding habits, semantic errors can be notoriously difficult to recognise.

In the next chapter we will consider the types of artefacts which are generated as part of a development process.

13 WORKING WITH SYSTEMS ANALYSIS ARTEFACTS

'What I hate about writing is the paperwork.'

– Peter De Vries

Although a working application is usually the end goal of developing a system, many different artefacts are created as by-products of the process, mostly in the form of documentation. Developers often dislike creating documentation, but many artefacts will naturally be created as part of the development flow. Common examples include class diagrams (schemas), project plans, business cases, risk assessments and user requirements.

In this chapter we'll examine two key examples of system analysis artefacts – the use case and user stories – and how these may be employed in the workplace.

USE CASES

A use case is essentially a list of steps or actions which define the interactions between a role and a system in order to achieve a goal of some kind. It is a technique for capturing functional requirements of a system for use in software development projects. Figure 13.1 shows an example.

Figure 13.1 Sample use case

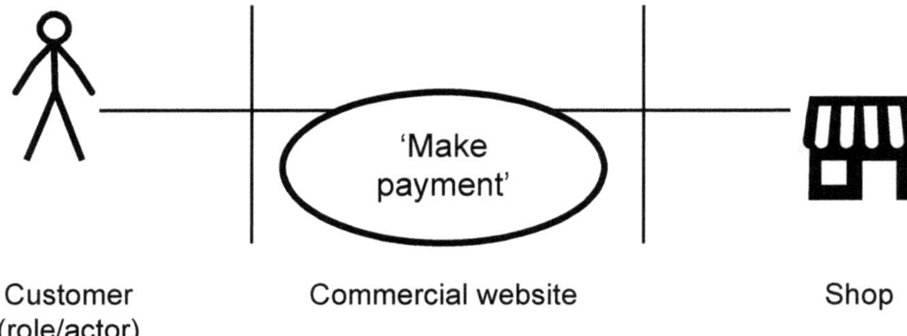

Use cases come in two basic flavours: system and business. The distinction is relatively straightforward:

- **Business use cases** generally describe at a fairly abstract level how a customer manually interacts with a business and the services they require, such as buying an item, making a payment or asking for an account balance.
- **System use cases** are used to model the lower-level interactions between the role and the various parts of the system. These are typically automated.

Note that terminology can vary across different analysis frameworks. For example, a 'role' is known as an 'actor' in UML (Unified Modelling Language). See Chapter 11 for more on UML.

There is also a relationship between 'actors' and stakeholders in a project. Actors are always stakeholders, but the inverse isn't always true. For example, a managing director who owns the projects may not have day-to-day interaction with the system, so they are a stakeholder but **not** an actor.

Use cases typically describe the basic flow of actions needed to perform a certain task and, if necessary, alternative flows if pre-conditions are not met. For example, a customer cannot make a payment if they haven't got any outstanding invoices; although a basic flow would support the user's request to make a payment, an alternative flow would need to indicate that payment is not required.

At no point in the use case should there be any implementation-specific language (i.e. no programming terminology). This keeps the solution programming language agnostic at this stage, because a target language may not yet have been chosen.

User stories are considered one of the most important artefacts generated from Agile frameworks such as Scrum. As such, in the next section we'll discuss them as part of the larger operational flow that a development team might encounter.

AGILE FRAMEWORKS IN PRACTICAL SOFTWARE DEVELOPMENT

As discussed earlier in this book, it is likely that you will encounter an Agile approach to software development within the modern workplace. This method advocates an iterative approach where customer requirements and the related solutions evolve over time, resulting in working software that closely meets the customer's needs.

Scrum is perhaps the most common Agile framework that you will encounter, although many others exist, including Kanban, rapid application development (RAD) and extreme programming (XP).

A key aspect of Scrum is the acceptance that a customer's needs will *undoubtedly* change over time, meaning that following a concrete-like 'fixed' plan which cannot be easily adapted is not productive. An Agile framework places more emphasis on the team's ability to adapt to new requirements and deliver these as quickly (and reliably) as possible.

In Scrum, a small team (typically three to nine members but this can vary) breaks work into several time-bound iterations called 'sprints'. These sprints commonly occur over a two-week period but can be longer or shorter. Progress on each sprint is tracked and reviewed in short 15-minute 'stand-up' meetings which are called 'daily scrums'; all members of the team typically contribute to these.

A sprint planning event, often taking up to four hours and led by the Scrum master, defines a sprint backlog from the product backlog, identifying items that can be successfully completed in one sprint (see Figure 13.2). Each item is then decomposed into detailed units of work known as 'tasks'.

Figure 13.2 Agile Scrum framework

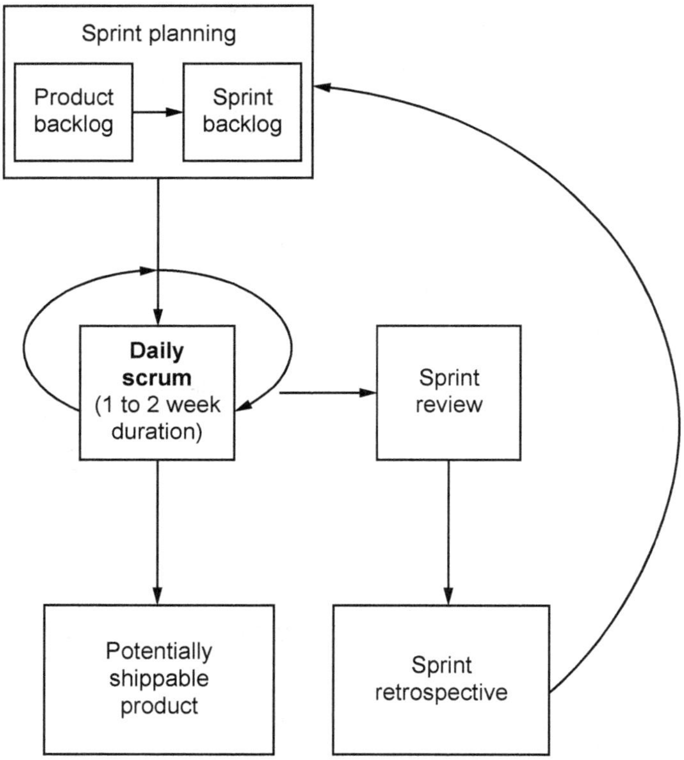

At the end of each sprint, a short two-hour review occurs where the team:

- reviews completed work;
- reviews planned work that was not completed;
- presents work to the product stakeholders (often in the form of a live demonstration).

User stories

A user story is a simple description of a desired software feature and it is written from the end user's perspective. Typically, it includes the following elements:

- a **role** (**who** wants it)
- a **feature** (**what** it is)
- a **reason** (and **why**!)

Figure 13.3 shows an example.

Figure 13.3 Sample user stories

User story #100

As an **administrator (role)** I need to be able to **change an employee's current team (feature), so that the email mailgroups are accurate (reason).**

User story #102

As a **helpdesk agent (role)** I need to be able to **see records of previous calls made (feature), so that I am aware of the customer's conversations with us (reason).**

User story #101

In my role as a **manager (role)** I must be able to **remove a user from the system once they have left the business (feature), so that security is maintained (reason).**

Ideally the user story should be so concise that it can be written on a small sticky note. In turn, the user story can typically be broken down into a number of discrete (but related) tasks. It is the development team's responsibility to create coded solutions that satisfy the requirements of each user story. In an ideal Agile environment, the programmers will consult with the end users to clarify any potential misunderstandings that might occur during the development process – you can't solve a problem if you truly don't understand it!

Agile task board

The task board is a visual tool used to show the progress made by a Scrum team during any particular sprint. Each sprint is composed of several different user stories and each story is represented by a number of tasks.

The board should allow team members to quickly see:

- tasks that haven't yet been started;
- tasks which are in progress;
- tasks which have been completed.

The task board may be a physical object, or it can be digitally available via an online interface, typically in the form of a cloud-based application such as Jira or Taigia.io.

Figure 13.4 shows a typical layout for a task board.

Figure 13.4 Agile task board

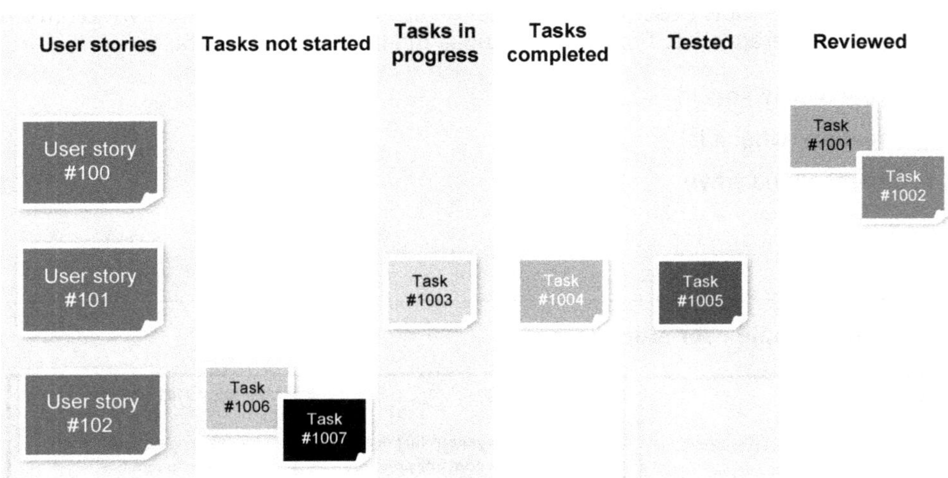

Each user story's associated tasks are initially placed in the 'Tasks not started' column. As the team's work progresses during the sprint, the status is changed by physically moving the tasks from left to right. The rate of progress on different tasks is typically not even (due to size, complexity or programmer throughput). This allows the Scrum master to quickly identify tasks which may have an associated 'impediment' or, worse still, a 'blocker', and see whether intervention (peer help or escalation to a more experienced programmer) is required.

An **impediment** slows down the pace of the team (think of it like a temporary obstacle that needs to be tackled) and may create waste (such as delays, defects or extra processes which the customer will not pay for). However, it ultimately won't stop the delivery of the product.

A **blocker** is more serious and may represent something that could prevent the actual delivery of the product. For example, a required software library might be updated and become incompatible with the core project. Blockers must be removed for the project to advance successfully. It is common for programmers to actively swap between active tasks while blockers are being escalated and resolved by more senior developers – this means that multitasking is a key skill for any modern programmer.

In the example shown in Figure 13.4, User Story #100 has been completed, with all tasks being successfully tested and reviewed. In contrast, User Story #101 has tasks in various stages of development but work on User Story #102 has not yet begun – this may indicate an impediment (such as a technical difficulty) or a resourcing issue, and hopefully not a blocker!

Although not shown here, it is common to see the assigned programmer's name tagged to each task along with the number of associated story points (see below).

Put simply, the Agile task board is a team's best way to track and communicate sprint progress in a transparent way to interested stakeholders. The daily scrum may typically occur in front of the task board, helping to keep everyone motivated and focused on completing the overall sprint goal. For a Scrum master, it is an essential tracker and review tool.

Story points

In Agile, points are assigned to each user story, typically reflecting the amount of effort it will take for a programmer to develop a given feature. The number of points does not reflect the underlying value of the product backlog item being worked on.

The number of points awarded takes into consideration the perceived complexity of the task, the perceived uncertainty and the degree of risk involved. In addition, story points may be revisited once calculated – for example, if the members of a development team change.

It is quite common to use a Fibonacci-style sequence when assigning story points. This means that a user story could be assigned 0, 1, 2, 3, 5, 8, 13 or 20 (and so on) points depending on the relative effort that has been estimated to be needed to solve the problem.

Depending on the Scrum implementation, points may be assigned by the Scrum master or be derived by estimates suggested by a range of team members – this technique is often called 'planning poker'. Differences in estimates should ideally be discussed rather than averaged.

A team may also set an upper limit for points per user story. If an item has an estimated number of points that is above this figure, it may suggest that the problem is too complex and really needs to be decomposed into smaller tasks.

Estimation is a process that improves with practice. Reviewing the accuracy of previous estimates (story points vs actual complexity and time taken) will help to improve future calculations, particularly when it is proven that an estimate was wildly optimistic and development took much longer than expected. This represents a good learning opportunity and helps the Scrum master to get a better feeling for the sprint velocity of their team (i.e. how much work they can complete in a single sprint).

SUMMARY

In this chapter we have reviewed some of the key system analysis artefacts that have a direct day-to-day bearing on the tasks completed by developers.

In the next chapter we will examine the skills and techniques used to manage and deploy code, taking note of the tools, changing approaches and practices involved.

14 BUILDING, MANAGING AND DEPLOYING CODE INTO ENTERPRISE ENVIRONMENTS

'The most powerful tool we have as developers is automation.'

– Scott Hanselman

As mentioned previously, few mid- to large-scale projects are completed by individuals in the modern software development industry – it's just become far too complex to work in this way. As such, developments need to be undertaken in environments where multiple individuals can contribute productively and collaboratively.

Creating a modern software development process involves adopting certain approaches, tools and techniques. In this chapter we will look at some of the more recent developments. But first, a bit of history.

PREVIOUSLY IN SOFTWARE DEVELOPMENT...

Many years ago, the different teams involved in the development of a new software project could be quite isolated and rarely interacted, instead focusing on their own jobs and responsibilities. For example, developers rarely talked to quality assurance (doing the testing) and didn't tend to interact a great deal with operations (who eventually deployed and monitored the solution) – that is, until things went wrong, of course.

As you might have guessed, this isn't a great model. There's no teamwork, no collective responsibility, and precious little communication or co-operation.

The solution to this is to modify your techniques by working together more effectively.

DEVOPS

As the name suggests, DevOps is the operational combination of 'developer' and 'operations', plus quality assurance (see Figure 14.1).

By improving communication through collaborative working in DevOps, the natural outcome is better-quality software, written more quickly and working more reliably. Toolchains are standardised and the focus is firmly placed on constant testing and continuous delivery, ideally meeting the requirements of an Agile delivery framework.

There are other advantages to DevOps, such as faster deployments and lower costs (bugs tend to be identified and fixed earlier in the cycle, and teams generally engage more

Figure 14.1 DevOps exists at the heart of three disciplinary areas

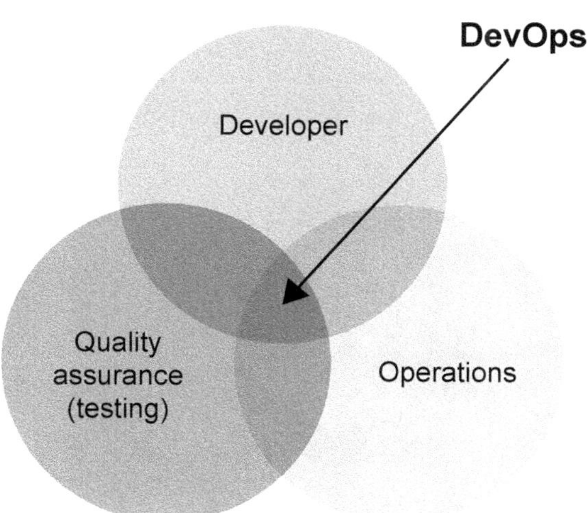

with the process, which improves underlying productivity). DevOps also encourages three adjacent practices:

- **Continuous integration:** developers merge their code into the shared repository more often.
- **Continuous development:** software is produced in shorter cycles, permitting more frequent releases.
- **Continuous deployment:** as above, but releases are made automatically through an automated production pipeline.

But still, something is missing: security.

DEVSECOPS

In the 21st century, it's no longer good enough to develop a solution that solves a set problem – it must also be secure. Simply put, this is because cyberattacks occur frequently at the application layer.

This is mostly because applications can be soft targets (they can be poorly programmed or use compromised components). This provides an attacker with a useful attack vector that they can use as an entry point before quickly escalating their privileges (e.g. by compromising user accounts, accessing databases or changing file permissions) on a now vulnerable infrastructure.

Obviously, vulnerable software that is easily exploited has exposed critical infrastructure, leading to worrying implications for data protection (GDPR; see Chapter 11) and financial implications for the organisation concerned (commercial and reputational damage).

The problem generally with DevOps is that traditional security models don't work particularly well. This has led to the concept of 'security as code' and the introduction of an even more tightly interwoven model of software production: DevSecOps (see Figure 14.2).

Figure 14.2 DevSecOps exists at the heart of four disciplinary areas

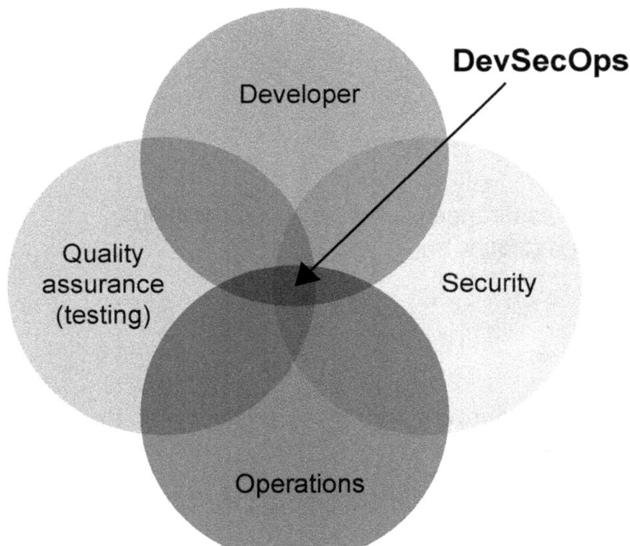

In this model, security is ever present throughout the development pipeline, being 'shifted to the left' (see Figures 15.1 and 15.2) rather than relegated to a pre-release penetration test. This allows security issues and common vulnerabilities (insecure libraries, embedded credentials etc.) to be identified and fixed early in the development process, further reducing costs and ensuring a more secure product. Automated pipelines are again the favoured applications used to manage this process.

SOFTWARE VERSIONING

In industry it is rare, but not unknown, for programmers to work in isolation on a software project. When they do, maintaining revisions of their program code is not an overly difficult task – they are completely aware of all the changes they make to their closed codebase. However, when programmers work in parallel on a project, keeping track of different contributions can be problematic; a version control system (VCS) is needed. When you start working as a developer, it is highly likely you will encounter one of these, so having some familiarity with them beforehand is a good idea.

Many popular VCSs exist (e.g. Concurrent Versions System, Subversion and Perforce), but perhaps the most ubiquitous is called Git.

Git was created in 2005 by Linus Torvalds as he worked on the Linux kernel with other programmers. It has subsequently exploded in popularity and become the preferred workflow-management tool for many organisations (and individual programmers) worldwide. Git is free and open source software, distributed under the GNU General Public License (GNU GPL) version 2.

Before you start thinking about Git, it is worthwhile remembering that projects are typically organised into a project structure, one that groups together related aspects of the solution. Understanding the structure of your project will enable you to quickly navigate its components when using a versioning system such as Git. An example project structure is shown in Figure 14.3.

Figure 14.3 A simple web project's structure

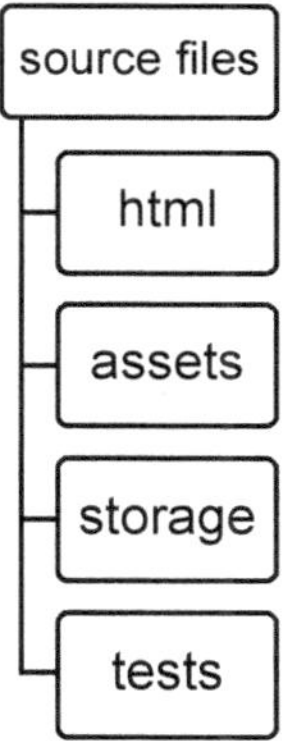

Git is freely downloadable and can be used to create a new local repository (containing your project) or clone from a public or private remote repository, although for the latter you need appropriate credentials.

How Git works

Unlike many other versioning systems, Git uses a 'snapshot' technique rather than keeping track of individual modifications in source code files. The snapshot contains state information about a folder.

The three areas of a Git project are shown in Figure 14.4. These areas all exist on your local computer.

Figure 14.4 Git projects

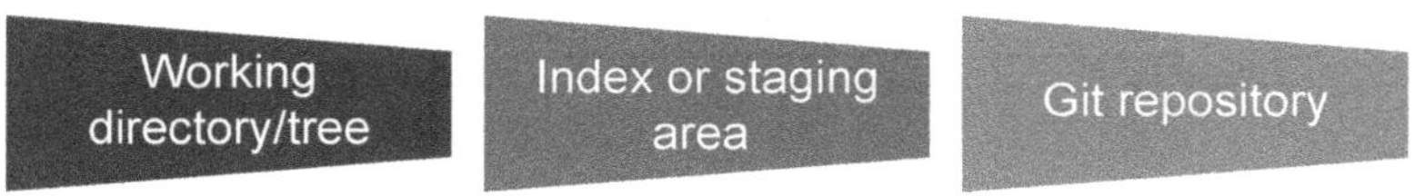

The working directory, or tree, usually starts in a completely unmodified state; this is often referred to as having a 'clean working tree'. As you add, delete or modify code in your source files, the tree state changes. Git considers all files to be either 'tracked' or 'untracked':

- **tracked file:** a file that Git knows about;
- **untracked file:** a file that Git does not know about.

A **`git add`** command is used to begin tracking a new file, effectively moving the modified file into the 'staging area' (it is now ready to be committed).

Unsurprisingly, a **`git commit`** command is used to commit your tracked changes to the local Git repository (any untracked changes you have are not included). It is considered good practice to include a **`Git comment`** which describes the nature of the commit you have made – a short but meaningful note can help project managers to track changes and identify the purpose of the commit.

At this point the required changes have only been made to your local Git repository. Eventually, a **`git push`** command can be used to 'share' your changes with the remote repository (often called the 'origin'). This is known as pushing your work 'upstream'.

Sometimes your push may be rejected because another programmer has pushed upstream just before you. When this happens, you must issue a **`git fetch`** or **`git pull`** to combine their changes with yours and then try again. Figure 14.5 summarises these processes.

Figure 14.5 Git repositories

Local repository			Remote repository
Working directory/tree	Index or staging area	Local branch	Remote branch
git add →			
	git commit →		
		git push →	
		← git pull	
	← git checkout		

Other programmers can then pull your changes from the remote repository down to their local git repository and see the changes you have made.

Branches

Developers often need to work on independent parts of a project which may not immediately impact others. To keep their work free of changes from other developers, they can work individually or collectively (in small teams) on separate 'branches' of a repository, rather than the traditional master branch.

Commits can be made to the new branch without affecting any other. When completed, each named branch can be merged back into the main codebase once it has been successfully tested (see Figure 14.6).

Figure 14.6 Branch example

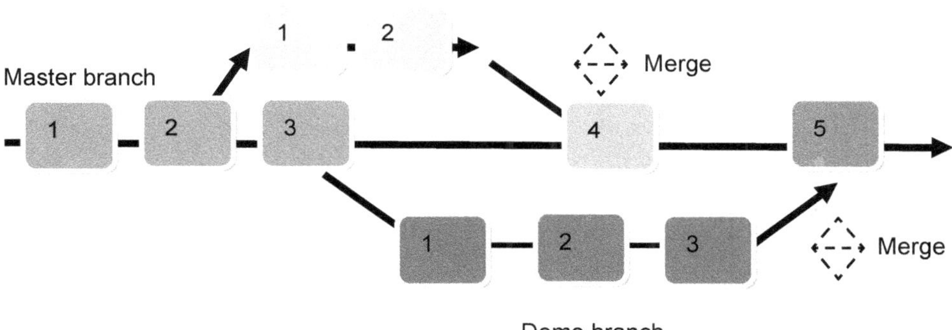

Branches are often used to test new ideas or experiment with problem-solving approaches without risking polluting the master branch, which is typically the one used in production.

Occasionally, having multiple branches can cause problems when the same source file has been modified separately by two programmers and their respective changes cannot be automatically reconciled by Git (typically they will each make different changes to the same code). When this happens, it is necessary for a programmer to perform a manual merge. This involves resolving flagged conflicts using an editor and then re-committing and pushing the combined changes back to the master branch on the remote repository.

Advantages of Git

Advantages of Git include:

- It provides a **long-term history** of every single file in a system which contains every change ever made. This includes:
  - file creation (including renaming and moving in the project structure);
  - file modification;
  - file deletion (including renaming and moving in the project structure).

- It **provides accountability**, as tracking every code change to a software project enables managers to easily identify:
    - annotation of each change (why it was done);
    - date of each change (when it was done);
    - developer commit signature (who did it).

  This can provide a project manager with important information which helps them to attribute project workloads and resolve problems or bugs that have been introduced in a recent (or historical) commit.
- It has **good data integrity:** SHA1 (Secure Hash Algorithm 1) is a cryptographic hash which provides a checksum (a value derived arithmetically from data that uniquely identifies it) that makes it simple to detect corruption.
- Git is a **distributed VCS**. This means that programmers can work independently on a project and choose which aspects of their work are uploaded for others to see. This also allows work to continue offline, if required.
- **Problematic changes can be undone** by inverting the changes made in the identified commit and appending a new commit with the necessary fixes.
- It works well as **part of an Agile workflow**; Git branches can be thought of as Agile tasks.
- Branches can be **tested individually**.

Disadvantages of Git

Disadvantages of Git include:

- **Operation can vary** between different operating systems.
- There are **many different commands** to learn (although admittedly a developer could survive with a core of 10–20 that are frequently used).
- It can require **more steps** than other VCSs to achieve the same goals – for example, making a change to a project is simpler in Subversion than in Git.
- **Command usage can be a little inconsistent** and therefore harder to learn.
- Git's first-party **documentation can be a little impenetrable** for the novice developer (for example, see https://git-scm.com/docs/git-push).

Web-based hosting services for Git

Commercial organisations may choose to host their own remote repository using Git or use one of the many different web-based services available. GitHub (acquired by Microsoft in 2018) is probably the most popular, having over 28 million users. It hosts millions of public open source and privately owned software projects. Bitbucket (owned by Atlassian) offers a similar service, including free accounts and various commercial hosting plans and support for other VCSs, such as Mercurial.

A popular feature of web-based VCS services such as GitHub is 'forking'. You may encounter a fork in a project when the codebase is split, for example when a sub-team uses the current codebase as a starting point for another project which will typically not feed back into the main one.

Git plug-ins

Many web-based hosting services offer 'webhook' plug-ins for continuous integration and delivery of automation services such as Jenkins. Such webhooks can be used to trigger an automated project build whenever a programmer commits new code to a web-based hosting service. Similar webhooks exist for Agile project tracking software, such as Jira.

Useful examples of webhooks include:

- ensuring that the Jira task/issue reference is included in the commit message;
- warning programmers about pulling from a remote repository's branch that contains known issues.

It is this flexibility that makes using a VCS a must-have skill for developers and a desirable trait that recruiters and potential employers will almost certainly seek out.

CHANGING DEVELOPMENTAL PRACTICES

Getting development practices right has always been a difficult proposition, principally because of the digital soup of different technologies which are typically required to form a modern software solution. For example, a developer may have their code working perfectly on their local development set-up but when it is being tested by others, the lack of an installed library or an inaccessible database could bring the application's execution to a screeching halt. This can be problematic because testers may not have the technical expertise to fix this type of broken mess (nor feel it's their responsibility to do so).

It can get much worse. For instance, it may be discovered during deployment that a client's target environment is not compatible with key components of the solution, making further progress fraught with difficulties.

Fortunately, the use of virtualisation (virtual machines, or VMs) and various cloud-based services provides potential solutions to these development, management and deployment issues. Until recently the use of VMs proved to be the most popular option, but these are typically resource hungry as they require a platform to host another operating system. Containers, offering a far more lightweight and portable alternative, are becoming much more popular.

The role of containers in software development and implementation

Containers have quickly become an incredibly useful tool for software developers over the past few years, saving them both time and effort when setting up their development and test environments. Given their growing popularity, its highly likely you will encounter them in the workplace.

Let's start by tackling the key question: what is a container? In simple terms, a container is a package of software that contains code and its dependencies. Dependencies are things a program needs in order to run successfully – for example, system and run-time libraries, and environment settings.

A traditional approach to hosting another operating system is to deploy a VM. However, a container differs from a VM because it attempts to virtualise the operating system, not the actual underlying hardware – there's no attempt to create another machine, just an environment which the programmer can use to develop and test their code. This lightweight approach makes containers much more efficient to use, and they typically consume far less memory and backing storage than a VM.

Figure 14.7 differentiates the two approaches. In this example, multiple containers can run on the same container engine, sharing the same resources provided by the host operating system (such as binaries and libraries) and underlying hardware but running as isolated processes. In the contrasting VM solution, the hardware runs three separate guest operating systems, with each requiring its own configuration, duplicate resources and so on.

Figure 14.7 Containers versus virtual machines

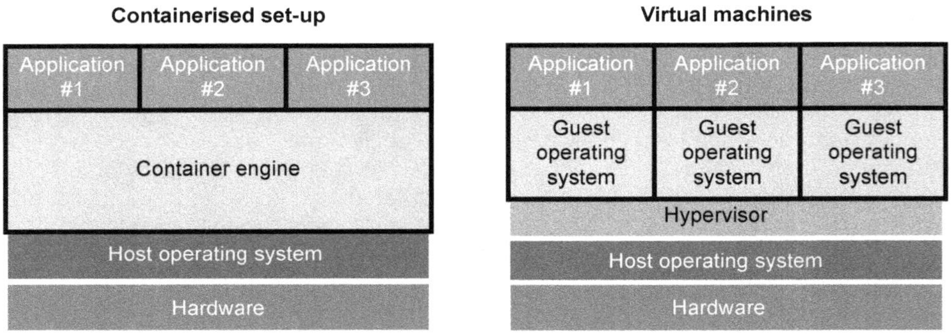

Probably the most well known **container engine** is Docker. It was introduced in 2013 and has subsequently helped to shape the industry standard through the donation of its container specification and run-time code to the Open Container Initiative.

Let's consider a practical, real-world example. If you had a development role as a full-stack web developer, it is likely that you would need to set up (at the very least) the following software:

- web server, such as Apache HTTPD;
- relational database server and client, such as MySQL or MariaDB.
- server-side script engine, such as PHP;
- server-side MVC (model-view-controller) framework, such as Laravel;
- IDE, such as IBM NetBeans.

That's at least five separate technologies which need to be installed and configured for use on a target operating system before any lines of code have been written. Of course, any variances between installation, configuration and usage in different operating systems, for instance Microsoft Windows vs Linux vs Apple macOS X, can only complicate matters further.

For example, case-sensitivity issues in some databases and programming languages are more important on Linux operating systems, which can cause nasty surprises for programmers developing on Microsoft Windows and then moving to a Linux staging environment for testing. Using a containerised solution for development makes things a lot easier as they tend to be Linux-based, so the issue typically does not occur – the programmer's development environment already matches the staging and (very likely) production environments where the code will eventually run.

A further example may be beneficial here. The following practical steps are required to quickly set up a suitable Python development container on a Windows PC:

1. **Install** Docker for Windows
2. **Pull** (download) an image of the container type required (e.g. MariaDB, Python or Ubuntu), typically from Docker Hub (hub.docker.com):

```
C:\Users\Demo>docker pull python
Using default tag: latest
latest: Pulling from library/python
22dbe790f715: Pull complete
0250231711a0: Pull complete
6fba9447437b: Pull complete
c2b4d327b352: Pull complete
270e1baa5299: Pull complete
8dc8edf0ab44: Pull complete
86ded05de41b: Pull complete
1eac5266a8fa: Pull complete
61b3f1392c29: Pull complete
Digest: sha256:166ad1e6ca19a4b84a8907d41c985549d89e80ceed2f7eafd90
dca3e328de23a
Status: Downloaded newer image for python:latest
```

3. Run the container (with suitable options) and start the Python shell:

```
C:\Users\Demo>docker run -it python bash
root@39867802043b:/# python
Python 3.7.2 (default, Mar 5 2019, 06:22:51)
[GCC 6.3.0 20170516] on linux
Type "help", "copyright", "credits" or "license" for more information.
>>> print("Hello World")
Hello World
>>>
```

4. You are now working within a Debian Linux container complete with an up-to-date version of Python. You can now start developing!

From a starting perspective, it is that simple. In practice, there's a lot of deeper learning required to get the most out of containerisation and the broad range of options and configurations that are available. Of course, it's also possible to build your own customised containers which you can share with others.

Deploying code into enterprise environments

One of the strongest aspects of the containerised approach is that it's possible to package your application (code, required libraries and dependencies) and its native environment into a container and 'ship' it to a customer, knowing it will run practically anywhere that has a compatible container engine. This has huge ramifications for implementation and the portability of solutions.

No wonder industry heavyweights such as Google, Microsoft and Oracle have embraced the technology; it's a game-changer.

PRODUCING A TECHNICAL GUIDE

All good development teams will produce a technical guide to support the development product. To ensure a technical guide is robust, the team should:

- Plan the content. The content should include some (or all) of the following, as appropriate:
  - description of the software;
  - systems requirements;
  - diagrams of the final solution (flowcharts, data flow diagrams, system diagrams, UI mockups, wireframes, storyboards etc.) (see Chapter 11);
  - annotated code;
  - diverse and representative test data.
- Carefully consider using the output from auto-documenters, which will almost always produce far more information that you really need to include. Some auto-documenters can produce hundreds of pages for what is essentially a relatively straightforward program or database.
- Ensure that correct spelling and grammar are used. This makes your documentation more authoritative and avoids it looking sloppy, rushed or unprofessional.
- Think about who is going to use the guide. Because this will be a technical guide, you are free to use technical terms, although you should still provide a glossary.
- Try to be concise and, where appropriate, use images to support the text – particularly flowcharts, other diagrams and tables of information – rather than just long paragraphs of text.

- Ensure code is commented appropriately. Comments in your coding contribute to technical documentation if not the technical guide itself.
- Ensure the technical guide is peer reviewed by other developers, both those involved in the development and others with suitable experience.

To help your client comply with the requirements of the GDPR, you should consider extending the product documentation to demonstrate how the product meets the requirements laid out in the regulations.

PRODUCING A USER GUIDE

A user guide is simply a document written to support people when they are operating your software. Most real-world appliances come with one, so why shouldn't your software?

Many users still prefer a physical user guide rather than an electronic one. This is because not every target user will have a dual-screen set-up, permitting simultaneous viewing of the guide and the software.

A user guide should include:

- an introduction including a concise description of the software's purpose;
- an overview of all the functionality within the software;
- information on how to install, configure and remove the software correctly (if applicable);
- screen captures with step-by-step instructions to walk users through operational procedures;
- if the product replaces a previous version, a 'before' and 'after' comparison of the user interface may be appropriate to help communicate the changes implemented;
- a trouble-shooting section including error messages that might be encountered (and their meanings);
- FAQs (frequently asked questions) are sometimes useful;
- a glossary of technical terms for quick reference.

Tips for creating a user guide

- Ensure that the guide is written using language that will be suitable for the experience level of the expected user (the target audience). Be consistent in what you call things and, above all, avoid technical jargon.

- Be careful not to patronise users. However, you should also not assume that your user has any real experience.
- Consider an electronic format (for environmental reasons).
- Depending on the deployment of your software, you may need to consider other natural languages (e.g. French, German, Urdu).
- Ensure that the content of your guide is logically organised.
- Think about the design of the pages and try to be consistent; familiarity helps people to absorb information.
- Ensure the guide is as simple as possible. You may think that cross-referencing will save on page count, but it can be frustrating for new users who must constantly flip backwards and forwards between different parts of the guide's content.
- Diagrams often help to explain complex ideas, and they certainly reduce the word count.
- Try to use ordered lists or bullet points rather than writing paragraphs of text.
- Use fonts consistently; many users prefer sans serif fonts such as Arial, Calibri or Helvetica because they are easier to read (particularly for on-screen documentation).
- User colour sparingly (apart from screenshots of the software, which benefit from being in colour).
- Ensure that there is sensible use of white space as a highly cluttered page can be frustrating for users.

TIPS FOR BUILDING, MANAGING AND DEPLOYING CODE INTO ENTERPRISE ENVIRONMENTS

Shown below are a few ideas for getting ready to work in a modern enterprise environment:

- Be familiar with VCSs: target popular ones such as Git and Subversion. You can always practise on personal projects by uploading these to remote repositories such as GitHub, where free accounts are available.
- Practise using VMs and containerised solutions to work in operating systems other than the host system you are familiar with (e.g. using a Linux container on a Windows PC to build and test software).
- Keep track of trends in software development processes (e.g. DevSecOps, continuous integration and continuous delivery).
- Remember that in a DevSecOps approach to software development you are likely to be communicating often with testers, hardware technicians and even network engineers; learn to speak their language.

SUMMARY

In this chapter we have seen that software development is rarely an individual activity in a modern enterprise, relying on a combination of innovative approaches, tools and techniques.

In the next chapter we will explore modern trends in software testing.

15 INDUSTRY APPROACHES TO TESTING

> 'Software never was perfect and won't get perfect. But is that a licence to create garbage? The missing ingredient is our reluctance to quantify quality.'
>
> – Boris Beizer

In this chapter we examine modern trends in software testing.

AUTOMATED TOOLS

Testing is often a time-intensive process and developers need feedback quickly in order to remediate problems with their code. The rise of automated tools makes this process less painful. Of the many available, common automated tools include IBM Rational Functional Tester, Ranorex, Selenium and TestComplete.

It is beyond the scope of this book to describe all those available and it should be remembered that each tool typically supports multiple programming languages and application types (such as desktop, mobile and web). However, some are particularly geared to certain types of application – for example, Selenium is very often seen as the go-to automation tool used for commercial web applications.

TRENDS

There are many trends in the software development sector that affect a team's approach to testing their code. Two in particular stand out: test-driven development (TDD) and secure (defensive) coding.

Test-driven development

This form of development practice is presently in vogue and often seen in software development job adverts as a desirable quality in a potential candidate. Rather than creating tests to check existing code, tests are written **first** (from use cases or user stories) and this shapes how code is then written.

Developers then write new code to pass these tests. Of course, the code may need further refactoring (see Chapter 7), but teams are assured that the test has been successful and therefore the component must have satisfactorily met the design requirements.

In TDD, each component is small and the development cycle is incredibly short, resulting in incremental – but reliable – development where less time is spent on reworking code and debugging.

Secure (defensive) coding

Due to the rise in popularity of network-aware and mobile applications, any modern testing strategy should place a considerable emphasis on security. Devices and their confidential data are obvious targets for cyberattacks, and poorly written applications are often the attack vector used by hackers to gain access.

Although commercial programming has been around for decades, it is only relatively recently that a concerted emphasis has been placed on application security (often abbreviated to 'AppSec' in the industry), primarily due to the software-related breaches which have plagued businesses across the world. Weak AppSec is one of the root causes of data breaches and may lead to reduced customer faith, very large fines, legal action and a level of reputational damage from which few organisations truly recover. AppSec is a feature of DevSecOps (see Chapter 14).

The goals of AppSec are to:

- identify potential threats (internal or external);
- protect critical data;
- fix existing security vulnerabilities;
- prevent future exploits from occurring.

Of course, from a software perspective, these tasks are particularly complex because vulnerabilities can occur in several different application types:

- cloud-based applications;
- desktop applications;
- web applications;
- mobile applications.

Each of these has its own technology challenges and distinctive set of vulnerabilities. As a developer it is important to know about these weaknesses and code accordingly. Table 15.1 lists some common vulnerabilities and their mitigations.

Table 15.1 Common security issues for software developers

Issue type	Description	Mitigation (how to resolve it)
Buffer overflow	A buffer is a temporary memory store, typically connected to an input stream (keyboard) or output stream (screen). Many languages, e.g. C (most classically), do not provide bounds checking, which allows much larger data values (typically strings) to be input into smaller storage locations on the stack. When this occurs, the buffer overflows into adjacent memory locations, which can corrupt other data items, potentially to the malicious user's advantage, e.g. by installing executable code.	Modern compilers often flag the use of vulnerable functions (i.e. library functions which are deemed insecure due to this potential for abuse). Typically, the functions are initially deprecated (usage is disapproved of) and then eventually may be physically removed from the library. This risks unsuccessful re-compilation of older code but forces the programmer to use more secure options instead. C's removal of the `gets()` function is a good example of this.

(Continued)

Table 15.1 (Continued)

Issue type	Description	Mitigation (how to resolve it)
Injection attacks	There are many different types of injection attack. One of the most common is the SQL injection attack, which occurs when malicious code is introduced into an executable SQL statement, typically through an unvalidated user input. This code then bypasses security, alters back-end databases and can be used to dump secure information to the user's screen or escalate privileges. Despite being flagged for years by OWASP (Open Web Application Security Project) and being preventable in most modern code frameworks, it still plagues modern web applications; we can only assume this is because of poor coding techniques.	The most common mitigation technique is to use parameterised queries which 'bind' values into pre-prepared SQL statements. These bound values, if injected with malicious code, typically fail to execute as they are syntactically incorrect; a bound value can only contain a particular type of value such as an integer or Boolean, not a full expression. Other useful techniques include use of pattern checking, cleaning the inputted string and limiting the permissions on sensitive database operations to particular users. Object-relational mapping techniques often offer suitable data-cleaning methods which can also remediate this potential weakness – although exploitation is still possible.
Insecure libraries	Many libraries have issues – i.e. vulnerabilities which can easily be exploited by malicious users. For example, the following once well-respected string-related functions in C are often targets of attack and should not be used: `gets()`, `strcpy()`, `strcat()` and `strcmp()`.	Use other functions which are less vulnerable, such as `strncpy()` (or `strlcpy()` if available). Some automated deployment tools, such as Jenkins, have 'dependency check' plug-ins which can scan for the use of insecure library functions in submitted source code before it is committed to a remote repository.
File opening	A common technique, particularly in web applications, is file path (or directory) traversal. This occurs when a file's pathname is modified to 'escape' a permitted directory on a server to target another vulnerable one, which may contain sensitive data, configuration files, source code and so on.	This is a simple one – never let the user specify the name of a data file to open! Either hardcode the string or allow the user to select from an available list of filenames, obfuscating the physical pathnames.

HOW SECURITY AFFECTS TESTING IN THE MODERN IT INDUSTRY

Historically, security has been relegated to the end of the production pipeline. At this point, whatever security issues exist in the code are likely to be 'baked in' – hard to isolate and consequently difficult to fix (see Figure 15.1).

Figure 15.1 Traditional approach to testing

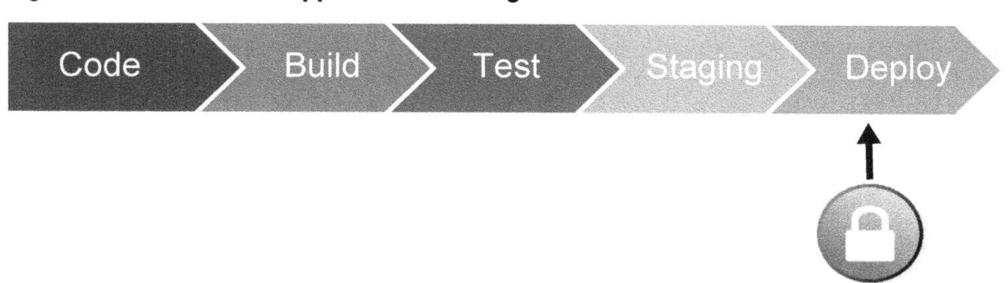

A popular term in DevSecOps (see Chapter 14) that emerged around late 2016 is 'shifting security to the left' (see Figure 15.2). More accurately, nowadays we can say we are shifting it to the very start of the process, rather than leaving it to the end of the production pipeline (at deployment).

Figure 15.2 Current thinking around testing

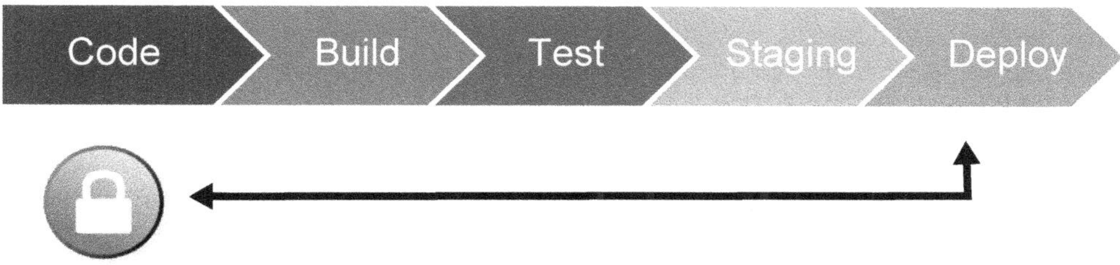

The effect of shifting security to the left is that vulnerabilities are discovered as early as possible in the production pipeline. These identified issues are typically resolved faster, often involving fewer people, which in turn saves the organisation both time and money. This represents a good return on investment for the organisation; the millions that may have been lost due to preventable data breaches in a deployed application may be reduced to mere hundreds of pounds in earlier security testing when a code defect or vulnerability is flagged and fixed – definitely a worthwhile investment!

The impact of continuous integration and continuous delivery on testing

Agile workflows which incorporate continuous integration and continuous delivery by their very nature strongly encourage the shifting of security to a much earlier position in the production pipeline. Generally, any new additions to the codebase should successfully pass a batch of tests before deployment. The recommended order is usually defined as follows (see Chapter 10 for more about these test types):

1. unit tests;
2. integration tests;
3. regression tests;
4. user tests.

It should be noted that creating automated testing within a pipeline is not an easy task. It often requires a high level of expertise (and commercial experience) with testing tools and the integration of these by, for example, launching them and capturing their outcomes. Developers should be intrinsically involved in this process, realising that testing is a critical process that should ideally be performed upon every single commit of code to a version-controlled repository.

TIPS FOR EFFECTIVE TESTING

Testing requires practice but there are some quick wins:

- Think about the extreme limits of your application's input (wherever they come from). What unlikely-to-happen ('edge') cases should you consider for testing?
- Have practical experience of a range of automated testing tools.
- Know your code's potential vulnerabilities – be aware of insecure libraries and common techniques (such as SQL injection) which are used to infiltrate applications and systems.
- Test early **and** often – move security aspects 'to the left' of the production pipeline. It is more effective and generally less costly to fix problems earlier, and it makes identifying and fixing code defects much faster.
- Be familiar with TDD.
- Cloud-based technologies bring fresh challenges for testing – be aware of platform-specific techniques.

Although acceptance tests that validate the software application against the identified business requirements require successful engagement with stakeholder end users who are familiar with the project's goals, these tests should not represent the first substantive engagement with the customer. Indeed, far from it!

SUMMARY

In this chapter we examined modern trends in testing and how security affects testing in the modern IT industry.

In the next chapter we will examine why developers need to stay focused on the needs of the client and stakeholders throughout the entire development process.

16 CLIENT AND STAKEHOLDER FOCUS

'Your most unhappy customers are your greatest source of learning.'

– Bill Gates

Successful software development relies on creating solutions that meet the needs of your client. Their role and level of day-to-day involvement in an ongoing project can vary extraordinarily, but their expectations are rarely set below perfection. Reality is often a more sobering proposition, so managing these expectations is vitally important. This chapter focuses on the client and wider stakeholders and how, as a developer, your relationships and interactions with them can shape a solution.

BEFORE DEVELOPMENT BEGINS

It is important to remember that any software development is a creative process, and everyone is aiming to create the best possible product. However, it is also a commercial endeavour and that means making certain agreements before development truly starts. What follows is a list of typical agreements that apply regardless of the type of application being developed:

- Agree the goal of the project – what does the client actually want to achieve?
- Agree the scope of the project – what is **and is not** within the project's boundaries to solve (this prevents scope creep; see Chapter 6).
- Agree the budget (and what this covers).
- Agree the timeline (when important milestones will be met and deliverables made available).
- Agree how communication will occur during and after the project: who, how (email, telephone, in person), when (timings, frequency) and so on.

SOFTWARE DEVELOPMENT CLIENTS

All software development projects have a client – this could be an entire organisation, a department within an organisation, or even a group or an individual end user of the product that is being developed. Clients can be internal or external.

Internal clients are those inside your organisation who are impacted by the development project and who will interact with the product (the outcome of the development). This works in much the same way as the IT support function in your organisation. Every employee in the organisation is essentially a customer of this function or department.

In software development, every employee who will use the proposed new or improved system is also classified as a client. This means that their needs should be considered in the design and functionality of the new product.

External clients are customers of your organisation who will be paying for your services. It's likely that the development project will have been formally estimated and that the client will have agreed to the proposed costs, timeframe and expected deliverables.

It shouldn't make any difference whether the client is internal or external; their treatment by a development team should be equally professional.

CHANNELS OF COMMUNICATION WITH YOUR CLIENTS

Interacting with your clients is essential for a number of reasons. Firstly, your clients need reassurance that the development project is progressing. In addition, involving a client and users in development will often result in a more successful product, an easier transition to a new system, and a higher level of buy-in from users and teams. In this situation, users will feel that they have been part of the change process, rather than being the recipients of change which is forced on them.

However, knowing what to communicate and what not to communicate to the customers of a development project can be a bit of a minefield. For example, does the client really need to know that development staff will be unavailable for three days in week seven due to a training event? Probably not. However, would the client benefit from knowing that a particular component has been completed? Possibly, if only to reassure them that the project is progressing. Would you tell your client if there had been some sort of technical issue which might have a significant impact on milestones and deadlines? Absolutely. Having said that, in this case the development team should first consider how they will overcome the problems so that they can not only alert the client to the problem but also explain the resolution, or at least be able to show that they have tried to resolve the problem.

Of course, the client must also take some responsibility here. If they desire a good product, they must engage with the process – for example, providing feedback in an iterative process, or giving guidance on the prioritisation of tasks in the project backlog.

What about the timeframe? How often should developers communicate with the client? The answer is as often as is necessary to get the job done and to ensure that the client is kept feeling informed about the development project.

Internal developments can be communicated through the organisation's intranet, with a weekly or a bi-weekly update that employees can read. Communications surrounding developments for external clients will probably be more formal, with review meetings scheduled at regular intervals so that the client can be brought up to date. They could even consist of some informal meetings just to keep in touch, as needed.

The strategy for communicating with your client should be agreed between you and them at the start of the development project to manage the client's expectations. However, the development team should be prepared to provide additional information or adjust the communication intervals (lengthening or shortening them as appropriate) as required by the client.

You should ensure that:

- the client understands the key milestones of the project and is informed as each one is approached and passed;
- the client is not unnecessarily alarmed by stories of problems experienced with the development project – unless, of course, the development costs are likely to increase or the completion deadline is under threat;
- communications are at the right level and have an appropriate level of detail in relation to the client (e.g. a non-technical client will not want a technical update).

Communications with your client should take place throughout the lifecycle of the development project, and the client (or their representative) should be invited to contribute at the review stage.

SUMMARY

In this chapter we have explored the importance of cultivating and retaining good client and stakeholder focus throughout the lifetime of a project and the benefits that this can bring to a development team.

In the next chapter we will examine the options for professional recognition.

17 PROFESSIONAL RECOGNITION

> 'Life is growth. If we stop growing, technically and spiritually, we are as good as dead.'
>
> – Morihei Ueshiba

Being recognised as a software developer of note typically comes through involvement in successful projects (particularly with regard to their size, complexity and inherent technologies), reputation of the employer involved and your career progression. However, as already discussed, it is important not to stand still, and that is where active continuing professional development (CPD) is crucial.

THE NEED FOR CONTINUING PROFESSIONAL DEVELOPMENT

CPD is critical for anyone working in the IT sector, and perhaps particularly so for software developers as they often work at the bleeding edge of technological change.

In an ever-evolving workplace where languages, frameworks, services and systems are rapidly being introduced, modified and discontinued in the merest blink of an eye, it is important that you actively track your accumulated skills and document your knowledge and experience. This will help you to identify any gaps and target potential opportunities for learning.

Gaining some form of professional recognition, typically through the completion of training courses and technical certification, is the natural by-product of this activity. In addition, having an IT-related skills framework within which to operate and evolve is also very useful to chart your progress. In this chapter we'll examine both these themes.

SKILLS FRAMEWORK FOR THE INFORMATION AGE

As a would-be IT professional, you should gain familiarity with the Skills Framework for the Information Age (SFIA). SFIA defines skills and competencies that practitioners should possess to be effective in different IT and computing roles across the sector. It is also a useful tool for employers as it allows them to compare the skills they **have** in their existing workforce against the skills they **need** to complete technical projects and retain their competitive edge.

For software development, BCS's guidance suggests that practitioners will be involved in:

> The planning, designing, creation, amending, verification, testing and documentation of new and amended software components in order to deliver agreed value to stakeholders. The identification, creation and application of agreed software

> development and security standards and processes. Adopting and adapting software development lifecycle models based on the context of the work and selecting appropriately from predictive (plan-driven) approaches or adaptive (iterative/agile) approaches.
>
> (SFIA Foundation 2020)

SFIA's six skills categories are:

- strategy and architecture;
- change and transformation;
- development and implementation;
- delivery and operation;
- skills and quality;
- relationships and engagement.

SFIA also outlines seven levels of responsibility. Levels 2 to 7 do not reflect academic levels; they outline the expected levels of responsibility that individuals will have. Level 2 practitioners, for example, will be expected to carry out instructions given by more senior colleagues while those at Level 6 will be developing organisational policies and strategies, planning and leading the software development activities within their organisations.

CERTIFICATION PROGRAMMES RECOGNISED BY INDUSTRY

Obtaining a form of vendor certification can give you an edge in a sometimes crowded software developer market. Of course, it's no substitute for comprehensive 'battle-hardened' experiential learning and the successful development and deployment of multiple industry projects, but it can open doors in organisations where particular frameworks, languages or toolsets are favoured.

We'll start by exploring one of the industry's market leaders: Microsoft.

Microsoft Certified Associate

From the end of June 2020, Microsoft is retiring its popular and recognisable Microsoft Certified System Administrator (MCSA), Microsoft Certified Solutions Developer (MCSD) and Microsoft Certified Solutions Expert (MCSE) certifications. This is part of the industry giant's gear change to more role-oriented training. So, while the commercially valuable Microsoft Certified Associate (MCA) label remains, the underlying exams and certificates are structured a little differently.

The roles identified by Microsoft include developer, data engineer, DevOps engineer, solutions architect, security engineer, AI engineer, data scientist, administrator and functional consultant. However, if we narrow our scope to the developer role, these can further be categorised as:

- Azure Developer;
- Microsoft 365 Developer;
- Dynamics 365 Finance and Operations Apps Developer;
- Power Apps + Dynamics 365 Developer.

For further information, visit https://docs.microsoft.com/en-us/learn/certifications/roles/developer.

In each category, pathway progress is marked using three levels: Fundamentals, Associate and Expert.

Now let's examine the different categories and some sample certifications.

Cloud technologies

At the time of writing it is becoming clear to many employers in the IT sector that their employees' knowledge of cloud technologies needs improvement; this is particularly true for developers. This is mainly due to the incredibly fast shift from 'tin' (i.e. physical servers) to virtualised computing and service-oriented architecture, both of which have found a natural home in the cloud.

Consequently, the demand for certification in cloud technologies has become a key concern, although due to market fragmentation there are a number of vendors involved.

Microsoft Azure Developer

For those interested in Microsoft's cloud solution, this pathway covers the technical aspects that a full-stack developer or software engineer needs to know. Typical topics include Azure App Service, functions, Cosmos DB (a schema-agnostic database), global apps (PaaS), search techniques, function triggers, availability and resistance, design and implementation, and performance.

Amazon Web Services Certified Developer – Associate Level

For those more interested in Amazon's cloud solution, this certification covers the broad range of skills required to develop, debug and deploy applications on Amazon Web Services (AWS). Typical topics include Elastic Compute Cloud (EC2), storing data and retrieving it from S3 (Amazon's simple storage service), DynamoDB (Amazon's relational database service), routing, Lambda functions for serverless applications, building APIs, monitoring application performance using CloudWatch and X-Ray, Identity and Access Management (IAM) authentication and authorisation, messaging, notification, caching and creating resources.

It is worth noting that Oracle also has certification for its cloud-based technologies, focusing on the overall instruction, SaaS and PaaS.

Language specific

Many programming languages have recognised certification programmes, whether these are first-party (i.e. provided by the language's vendor) or third-party 'vendor-neutral' formats made available through independent organisations. There are far too many to cover here, but representative examples include the following.

OpenEDG Python Institute certifications

The Python Institute is an independent not-for-profit project set up by the Open Education and Development Group. Its aims are threefold:

1. to promote the language;
2. to train a new generation of Python developers;
3. to support professional careers in the language and its related technologies.

The certification aims to provide a global standard in Python programming for software developers, providing recognition of skills, knowledge and overall coding proficiency in the language. It is supported by Pearson VUE (a world leader in computer-based testing). The examinations map content to three competency levels:

- Certified Entry-Level Python Programmer (PCEP) – bronze level;
- Certified Associate in Python Programming (PCAP) – silver level;
- Certified Professional in Python Programming (PCPP) – gold level.

For further information, visit https://pythoninstitute.org.

JavaScript certifications

A simple option is to consider the W3Schools JavaScript Developer Certificate. This represents an introductory validation of JavaScript and HTML document object model knowledge and skills.

For further information, visit https://www.w3schools.com/cert/cert_javascript.asp.

The Certified Internet Web Professional (CIW) programme is another consideration for JavaScript programmers as it has a series of qualifications targeting specific web development areas, for example:

- Web Foundation Series;
- Web and Mobile Design Series;
- Web Design Series;
- Web Development Series;
- Web Security Series.

Additionally, the Web Development Professional series is of particular interest as it contains a specific CIW JavaScript specialist certificate.

For further information, visit https://www.ciwcertified.com/ciw-certifications.

Oracle Java certifications

At the time of writing, Oracle University offers the popular Oracle Certified Professional (OCP) Java SE (Standard Edition) Developer credential, which has widespread industry recognition. Although this is primarily made available for the current version of the language, earlier versions remain available (and supported) for some time.

Other certification levels are available for various versions of the language, including:

- Oracle Certified Junior Associate – novice-level certification;
- Oracle Certified Associate (OCA) – SE-based certification;
- Oracle Certified Master (OCM) – SE-based certification;
- Oracle Certified Master (OCM) – EE (Enterprise Edition) certification.

For further information, visit https://www.oracle.com/uk/corporate/features/oracle-certification.html.

TIPS FOR GETTING PROFESSIONAL RECOGNITION

Here are some ideas for acquiring recognition and the reasons for doing so:

- Recognised frameworks provide a benchmarking mechanism.
- Value the role of CPD; choose an employer that invests in its employees and that values CPD to the same degree as you do.
- Identify your training needs and communicate these clearly to employers and other relevant people and organisations.
- Set yourself clear career goals and chart how these can be achieved.
- Identify what skills (and levels of experience) are required to function in a given role.
- Remember that certification isn't forever – it will require periodic updating.

SUMMARY

In this chapter we have explored the options for professional recognition, noting that this is a snapshot of the certification market when this book was written and so routes to professional recognition will change as software technologies advance.

18 FINAL THOUGHTS

```
while (!successful)
{
  successful = keepTrying();
}
```

HOW THINGS CHANGE...

Getting your first software development role is typically exciting, challenging and somewhat frightening in equal measures. You have graduated from creating code that is designed to be tested in classrooms for assessment purposes to crafting code which is live, customer-facing and critically important to the everyday function of a business entity. That's an incredible leap in responsibility.

Previously, if your code didn't work completely as requested, it wasn't generally disastrous. But in the real world, faulty code can result in serious financial errors, unexpected data loss, unwanted data breaches or far worse.

Who, then, would honestly want to be a software developer? Well, we hope you still do!

PRACTICE MAKES PERFECT

All software developers make mistakes and it is likely, should you start down this path, that you will be no different. Learn from your mistakes and never be afraid to ask if in doubt. Software development generally favours asking for permission rather than forgiveness. Also remember that the growth of technical skills is not the only goal – personal and interpersonal skills such as problem-solving, analysis, communication and time management are also critical to success and scaling your role upwards.

IDENTIFY YOUR OPPORTUNITIES

Higher positions may not always be found in the same organisation; be brave and embrace emerging technologies, and you will likely find new and exciting opportunities coming your way.

Your previous software development projects form the digital 'shop window' that potential employers will scrutinise to see whether you are a good fit for their development needs. How you present your ideas at interview will demonstrate the soft skills they are seeking to complement their existing development team.

Always strive to improve. Accept offers of training when provided and always make the most of them. Learn new techniques from other developers (young and old – don't discriminate) and always explore technology shifts and emerging trends – you never know when they will become the next big thing.

Above all, enjoy the challenge!

REFERENCES

Agile Manifesto Authors (2001) Manifesto for Agile Software Development. Available from: https://agilemanifesto.org.

Chonoles, Michael Jesse, and James A. Schardt (2013) *UML 2 for Dummies*. Hoboken: Wiley.

de Voil, Nick (2020) *User Experience Foundations*. Swindon: BCS.

Dowling, Lewis (2020) *Top 7 Programming Languages of 2020*. Available from: https://www.codingdojo.com/blog/top-7-programming-languages-of-2020.

Gladwell, Malcolm (2008) *Outliers: The Story of Success*. London: Penguin.

Gov.uk (2020) Software developer. Available from: https://www.gov.uk/guidance/software-developer.

Institute for Apprenticeships & Technical Education (2020) Software developer. Available from: https://www.instituteforapprenticeships.org/apprenticeship-standards/software-developer.

Intellectual Property Office (2016) Copyright notices. Available from: https://www.gov.uk/guidance/copyright-notices.

Kim, Moses (2018) Programmer vs developer vs engineer: the naming dispute. Available from: https://medium.com/shakuro/programmer-vs-developer-vs-engineer-91ef374e5033.

Kniberg, Henrik (2016) Making sense of MVP (minimum viable product) – and why I prefer earliest testable/usable/lovable. Available from: https://blog.crisp.se/2016/01/25/henrikkniberg/making-sense-of-mvp.

Maughan, Alistair (2010) Six reasons why the NHS National Programme for IT failed. *Computer Weekly*. Available from: https://www.computerweekly.com/opinion/Six-reasons-why-the-NHS-National-Programme-for-IT-failed.

Measey, Peter et al. (2015) *Agile Foundations*. Swindon: BCS.

NOS (National Occupational Standards) (n.d.) Software Development Level 3 Role. Available from: https://www.ukstandards.org.uk/PublishedNos-old/ESKITP5023.pdf.

Octoverse (2019) The state of the Octoverse. Available from: https://octoverse.github.com.

Osman, Maddy (2020) Wild and interesting WordPress statistics and facts (2020). Available from: https://kinsta.com/blog/wordpress-statistics.

OWASP (2017) OWASP top ten. Available from: https://owasp.org/www-project-top-ten.

SFIA Foundation (2020) Systems development. Available from: https://www.sfia-online.org/en/framework/sfia-7/skills/solution-development-and-implementation/systems-development.

SmartDraw (2017) What is a flowchart: flowchart symbols, flowchart types, and more. Available from: https://www.youtube.com/watch?v=iJmcgQRk048.

SmartDraw (2018a) All about UML activity diagrams. Available from: https://www.youtube.com/watch?v=Wf_xlagfHmg.

SmartDraw (2018b) Data flow diagrams: what is DFD? Data flow diagram symbols and more. Available from: https://www.youtube.com/watch?v=6VGTvgaJllM.

Stack Overflow (2019) Developer survey results. Available from: https://insights.stackoverflow.com/survey/2019.

Sutton, David (2019) *Cyber Security: A Practitioner's Guide*. Swindon: BCS.

UK National Careers Service (2020) Software developer. Available from: https://nationalcareers.service.gov.uk/job-profiles/software-developer.

W3C (2018) Web Content Accessibility Guidelines (WCAG) overview. Available from: https://www.w3.org/WAI/standards-guidelines/wcag.

FURTHER READING

It is impossible to cover all practical aspects of a role as broad and diverse as a software developer in a single book. However, we hope that your appetite has been whetted to learn more. Following are some further publications which you might find useful.

BOOKS

Clarke, Jill (2020) *Software Developer*. Swindon: BCS.

de Voil, Nick (2020) *User Experience Foundations*. Swindon: BCS.

European Union (2018) *General Data Protection Regulations*. Available from: https://gdpr-info.eu.

Gamma, Erich, Richard Helm, Ralph Johnson and John Vlissides (1994) *Design Patterns: Elements of Reusable Object-Oriented Software*. Boston: Addison Wesley.

Griffiths, Ian (2020) *Programming C# 8.0: Build Windows, Web and Desktop Applications*. Sebastopol, Calif.: O'Reilly.

Hambling, Brian (2020) *Software Testing*. Swindon: BCS.

Holmes, Lee (2012) *Windows PowerShell Pocket Reference*. Sebastopol, Calif.: O'Reilly.

Josuttis, Nicolai M. (2012) *The C++ Standard Library: A Tutorial and Reference*. Reading, Mass.: Addison Wesley.

Lutz, Mark (2013) *Learning Python* (5th ed.). Sebastopol, Calif.: O'Reilly.

Myers, Dominic (2020) *Front-End Developer*. Swindon: BCS.

Paul, Debra, and James Cadle (2014) *Business Analysis* (3rd ed.). Swindon: BCS.

Robbins, Arnold (2016) *Bash Pocket Reference*. Sebastopol, Calif.: O'Reilly.

Silverman, Richard E. (2013) *Git Pocket Guide*. Sebastopol, Calif.: O'Reilly.

Stroustrup, Bjarne (2013) *The C++ Programming Language*. Reading, Mass.: Addison Wesley.

Ward, Brian (2020) *How Linux Works* (3rd ed.). San Francisco: No Starch Press.

WEBSITES

Coding Ground, https://www.tutorialspoint.com/codingground.htm: a great resource for would-be software developers as it has versions of all of the main programming languages (C, C++, Python, Java, Ruby) which can be used through a browser, without the need to download anything

Common Vulnerabilities and Exposures (CVE), https://www.cvedetails.com

Open Source Resources, https://opensource.com/resources: IT-related articles in a number of different categories, including programming and DevOps

Open Web Application Security Project (OWASP), https://owasp.org

Python

Python Community, https://www.python.org/community

Python 3 Module of the Week, https://pymotw.com/3

Python 3 Patterns, Recipes and Idioms, https://python-3-patterns-idioms-test.readthedocs.io/en/latest/index.html

Learning development skills

Cloud Academy, http://cloudacademy.com

Code Academy, https://www.codecademy.com

Learning web technologies and best practice

Common Vulnerabilities and Exposures, http://cve.mitre.org

Httpbin.org (a simple HTTP request and response service), http://httpbin.org

Laravel: The PHP Framework for Web Artisans, https://laravel.com

Open Web Application Security Project (OWASP), https://owasp.org

Postman: The Collaboration Platform for API Development, https://www.postman.com

WebScraper Test Sites, https://webscraper.io/test-sites

Careers

Career Connect (advice service), https://www.careerconnect.org.uk

The Guardian (careers advice), https://jobs.theguardian.com/careers

National Careers Service (Software Developer), https://nationalcareers.service.gov.uk/job-profiles/software-developer

Next Step (government service for adults), https://www.gov.uk/government/news/careers-guidance-service-helping-adults-take-next-step

Prospects (careers advice), https://www.prospects.ac.uk/careers-advice

UCAS (careers advice), https://www.ucas.com/careers-advice

GLOSSARY

Agile: A project management methodology that is especially popular in software development projects but can also be used in other sectors.

Algorithm: A sequence of steps or operations that are created to solve a problem.

AppSec: Application security: the complex art of protecting software from being compromised by removing vulnerabilities that could be maliciously exploited.

Backlog: In the Agile methodology, a backlog is list of deliverables that need to be created for a given software development project.

Black box testing: Testing the functionality of a software product without considering (or having any knowledge of) the underlying code.

Changeover strategy: A planned process for changing from one way of doing something to another (e.g. moving from a paper-based accounting system to a computerised one).

Class: A software 'capsule', consisting of interrelated properties and methods that relate to a particular 'thing' that exists (e.g. a car, a person or a bank account).

Client-side: Where the actions take place on the user's own computer system (e.g. JavaScript running within a web browser).

Codebase: The collective term for the various files and lines of code in a computer program.

Commit: To confirm changes and make them permanent, particularly in version control; can be used as a verb or a noun.

Compilation: The process of translating source code into machine code for the CPU to execute.

Construct: A building block which is used in the development of a program. They are sequence, selection (choice) and iteration (loop).

Container: A data type which groups together multiple data items (e.g. an array, a queue or a stack).

Development environment: The working platform (hardware, operating system etc.) on which program code is initially written and tested by the software developer.

DevOps: The practical combination of software development, IT operations and quality assurance in order to produce high-quality software.

DevSecOps: The same as above, but also focusing on application security.

Encapsulation: The process of combining the attributes and methods of an object (which forms a class).

Enterprise glue: Programs or scripts that connect different parts of a system together to form a working data pipeline or process.

Environment tier: Different contexts where software is developed or used. For example, a sandbox has a full working version of a system, including its data, but is isolated from the live version currently available to users. This allows the developer to experiment with the components of the program without affecting the live version. See also **Development environment**; **Production environment**; **Staging environment**.

Evergreen: Something which is considered always relevant. For example, evergreen content on a website does not become dated.

Floating-point number: A number with a decimal part (e.g. 31.56).

Formative testing: Testing which is carried out on code while it is being developed.

Functional programming: A programming paradigm where data processing can be completed using many different functions in concert; a popular aspect of data-mining techniques.

Heap: Free memory which is allocated dynamically by a programmer at run-time and is larger than the memory afforded by a stack.

HTTP request: Part of the communication between a computer and a server which uses the Hypertext Transfer Protocol. A request is usually followed by a response.

Instantiation: The process of creating instances of a class as individual objects.

Integer: A whole number with no decimal part (e.g. 101).

Interoperability: Compatibility of a new system with other existing components, products or technologies. This is typically a result of different organisations working to shared standards.

Language: A set of computer instructions (such as C++, PHP, Python or Visual Basic) that are used to create applications, programs and apps. The syntax of languages invariably differs, although they have common constructs and parallels can often be found.

Library: A collection of functions and methods with a common theme (e.g. a maths library containing prewritten mathematical functions, or a string-handling library with functions to manipulate the string data type).

Linting: The process of checking for common coding problems (such as badly chosen variable names, inconsistent logic or the presence of unused variables).

Minimum viable product (MVP): An MVP will do a basic job, but it would benefit from refinements with additional functionality or the incorporation of emerging technologies.

Modularity: The process of dividing a problem into small, manageable chunks so that they can be developed individually and then merged at the end to form a software product.

Namespace: This term is used as a container for constants and variables. It defines the scope in which these items can be accessed.

Object-oriented programming: A software approach which is based on the concept of modelling the 'things' which interact in the real world. These things may be physical, such as people, or logical, such as a bank account.

Operator: A symbol (or symbols) which tells a compiler what to do to produce a result. A '+' will add two numbers together. A '<=' will check whether the number on the left of the symbols is less than or equal to the number on the right and will produce a result of true or false accordingly. Operators are sometimes overloaded to perform different actions on particular data types – for example, for strings, a '+' may mean concatenation.

Pair programming: An Agile software development technique which involves two programmers working together on a single problem at the same terminal.

Procedural programming: A programming paradigm where problems are solved using a defined set of instructions.

Production environment: A live system where a software solution is executed.

Requirements specification: A blueprint for a software development which sets out what the final product should do, often separated into 'must haves' and 'desirables'.

Run-time library: A set of routines which are needed when a program is run to manage the behaviours of the program in the executable environment.

Scrum: An Agile framework that is based on the acceptance that a customer's needs will change over time. This enables software developments to flow and change, rather than sticking to a rigidly defined plan. It results in short iteration cycles which deliver new/changed products as quickly as possible.

Server-side: Where the actions take place on a remote server rather than on the user's computer system (the opposite to client-side).

SQL injection: An aggressive hacking technique which exploits vulnerabilities in programming code by allowing malicious code to be run by modifying the purpose of existing SQL statements. Often occurs when HTML form-based inputs are not correctly sanitised before processing.

Stackframe: Data which gets pushed onto the stack when a function is called. It normally contains the actual function call, its arguments and local data.

Stacktrace: Sometimes called a backtrace, it is a reverse chronological listing of functions called at the point when a run-time error/exception occurs; it helps the software developer to determine the root cause of a problem.

Staging environment: A duplication of the hardware, operating system and so on that a software product is designed to be used in. Unlike a production ('live') system, it is not in the public domain and can therefore be used for realistic testing.

Summative testing: Testing which is carried out on code once development is complete to check it meets the client's requirements (among other things).

Superset: Enhanced or expanded features in a programming language.

Syntax: The written elements, structure and rules of a programming language.

Toolchain: A group of programming tools which are used together to perform a complex programming task; often includes the editor, compiler, linker, debugger and so on.

Unit testing: Testing small parts of a program (such as individual functions) using a white box testing approach to check the accuracy and performance of an individual component.

Virtualisation: The creation of a non-physical version of a computer system on a host, including its operating system and underlying hardware.

Webhook: An automated process which is executed in response to an event occurring on a website (e.g. a developer pushing their code to a remote repository or a customer submitting a complaint via an online form).

White box testing: The process of testing the logic of the program. This means that each pathway through the program (each selection and each loop) is individually tested for all possible routes to make sure that each works as intended. For example, it might test that every option on a menu executes when chosen and that the 'exit' option closes the program. The functionality is not relevant at this stage (that will have been tested in black box testing).

INDEX

www.ingramcontent.com/pod-product-compliance
Lightning Source LLC
Chambersburg PA
CBHW060551060326
40690CB00017B/3679
* 9 7 8 1 7 8 0 1 7 4 9 7 6 *